Disaster Studies and Management

Aims and Scope

This book series aims to further academic scholarship and discourse in the rapidly emerging discipline of disaster studies – encompassing both theoretical and practical concerns in the interface between disasters, their conception, their management and their relationship with various aspects of governance. The series will cover diverse perspectives and approaches, reflected in case studies and narratives from different parts of the world.

The series addresses the growing need to bring together the wealth of dialogue and experiences in disaster research and disaster management. It seeks to enhance the knowledge base explicating the complex relationship between shifting socio-economic situations, unplanned urbanization, environmental degradation, climate variability and change, geological hazards and the threat of epidemics. The series adopts both a comparative and a critical perspective influencing the contemporary discourse around disasters, development and the use of appropriate technologies to foster safety, security and sustainability of the planet.

The series is multidisciplinary in its orientation and invites contributions from academicians, policy makers, practitioners, consultants working in the broad field of disaster management desirous of making scholarly contributions. Books in this series would be of interest to students pursuing postgraduate courses and to practitioners and policy makers looking for reflective and critical literature related to disaster governance and development.

The Series:

- provides a comprehensive coverage of contemporary issues in disaster management and disaster studies
- contributes to developing theoretical rigour, linking the fields of disasters, development and climate change adaptation
- focuses on critical analysis and developing a comparative perspective in disaster management

Expected contents could be related to but not limited to:

✓ Theorization in disaster studies: Perspectives and ideological moorings
✓ Disaster risk reduction, disasters and development
✓ Critical review of evolution of DM practices in diverse contexts
✓ Disasters and media, crisis communication, disasters in a hyper-mediated world
✓ Narrative analysis, ethnography and anthropology in disaster studies
✓ Education, capacity building in DM
✓ Climate change adaptation, extreme weather events and disasters
✓ Cascading disasters or complex emergencies
✓ Disasters and conflict, involuntary displacement, migration and disasters
✓ Local and cultural knowledge for disaster preparedness and response
✓ Critical reflections/analysis on DM as an emerging profession
✓ Law, governance, human rights and transboundary governance
✓ Accountability, entitlements and ethics in DM
✓ Disasters, politics, humanitarian aid
✓ Globalisation, role of UN and international organizations in DM
✓ Technological approaches to DM, assessments and mapping
✓ Health, public health and hospital preparedness (water, sanitation, epidemics)
✓ DM as pluralist practice: Lessons learnt (pre and post disasters, before and after studies)
✓ Disaster "recovery" practices (economy, habitats, land use, housing, livelihoods, religious and cultural practices)
✓ Disasters, exclusion (special groups- caste, class, elderly, disabled, tourists, etc.)
✓ Data needs and decision making tools in DM, GIS—based approaches to DM
✓ Disaster logistics and supply chain management

Taken two years after the Indian Ocean Tsunami of 2004, the cover picture represents the sheer grit of Nicobari tribals, a fiercely independent community of Chaura, a low lying, tiny island in the Andaman and Nicobar Islands of India. Even as the administration evacuated them to temporary shelters on Teressa island for an indefinite period of time, the tribals, tired of waiting for government promises, built their own boats and moved back to their islands. An embarrassed administration was then forced to take note and provide livelihood and other forms of support.

Mohammad Tarikul Islam

Disaster, Governance and Development

Perspectives from Bangladesh

Springer

Mohammad Tarikul Islam
Department of Government and Politics
Jahangirnagar University
Dhaka, Bangladesh

ISSN 2730-5635 ISSN 2730-5643 (electronic)
Disaster Studies and Management
ISBN 978-981-99-1630-6 ISBN 978-981-99-1631-3 (eBook)
https://doi.org/10.1007/978-981-99-1631-3

This Springer imprint is published by the registered company Springer Nature Singapore Pte Ltd.
The registered company address is: 152 Beach Road, #21-01/04 Gateway East, Singapore 189721, Singapore

Foreword

In the age of the Anthropocene, where man-made changes to the physical environment and the crust of the earth now precipitate various forms of climate change, Bangladesh has been at the forefront of complex actions and efforts to alleviate the resulting human suffering and mass destruction. Bengal as a region of South Asia and Bangladesh as a nation have been no stranger to man-made atrocities, natural disasters, and numerous catastrophes linked to climate change. Bangladesh earned a surprising reputation with its efforts despite being labeled a "Basket Case" by Henry Kissinger shortly after gaining independence. Bangladesh has celebrated 50 years of independence in 2021. It has not only become one of the most populous countries in the world but is also a model of activism in regard to addressing the urgent requirements of physical and human development as well as disaster management. On my visits to Bangladesh, I saw first-hand not only that Dhaka never seems to sleep, and traffic rolls all night, but that many people work extremely hard, often in difficult and atrocious conditions, to secure survival, a decent living, and a safe future for themselves and their children. As natural disasters cannot be completely avoided and have in fact become part of the lived experience and history of this region of the world, they have to be faced individually and collectively, either by mitigating the already ongoing or occurring effects of natural disasters or adapting to them as skillfully as possible. There is no time for long debates anymore, as we now learn from studies that focus on the effects of the Anthropocene. Hands-on action has to be taken. Years ago, I saw for myself in coastal areas of Bangladesh how basic flood shelters now dot the landscape as a symbol of human efforts to mitigate suffering. But much more is evidently needed. It is crystal clear that various adaptation processes as a form of empowerment at all levels need to be adequately planned and executed. As the most drastic effects of climate change disasters hit directly at the local level, it makes perfect sense to conceptualize them as a crucial form of private–public partnership, utilizing and activating the existing bonds between citizens and the government at various levels. Every citizen needs to be involved in some form, as cyclones and floods do not discriminate on the basis of religion, ethnicity, or class. They simply ruin everything and everyone in their paths. While the state carries the overall responsibility for coordinating the various efforts, political differences and disagreements over all kinds of issues have to be put aside to engage in concerted actions of response.

Meanwhile, Bangladesh has become a global model for how to manage such challenges. As a result, the basic terminology of "disaster management" has not only become an important topic in various academic disciplines but has experienced a remarkable expansion and sophistication. This is clearly reflected in the core concept discussed in this important book, which is effective disaster management with transparency, accountability and in terms of sustainable development. Disaster risk reduction (DRR), which indicates an awareness that while natural disasters are inevitable, concerted action may and indeed must be taken to minimize risks. Obviously, this involves many stakeholders who need to learn to work together, often in new ways, to achieve maximum efficiency. The alert consciousness of DRR culture, which this study promotes, extends from the national level, through the National Plan for Disaster Management of 2007–2015, followed by the Disaster Management Act in 2012, right down to the local level. What appears and is discussed as most important here is the paradigm shift from the earlier post-disaster relief culture to the now prevailing DRR culture. This involves consideration of the opportunities as well as challenges for mainstreaming disaster management processes into integrated development structures through responsible forms of governance at all levels, starting from individual consciousness and self-help to various forms of collective and community-based processes and several layers of government structure, right to the top. Here, too, political leadership as visionary management is important.

Temptations to manipulate, exploit, or obstruct the necessary multi-dimensional activities involved will continue to exist. All along, activist efforts will need to focus on devising and experimenting with new and revised methods, as well as assessing achievements and failures. The challenges are truly enormous, but linking this whole field to development processes is further evidence that climate change also contributes to new forms of human development in various domains. The resulting interdisciplinary synergies will be obvious. The work of natural scientists, the engagement of human sciences, and various forms of governance structures as well as international solidarity efforts and their implications will need to be factored into a comprehensive analysis. Bangladesh has all of this in abundance and needs to make constructive use of its resources.

The author of this pioneering study is a well-prepared young political scientist, whose strong academic efforts as well as direct practical experience within the development sector make him ideally suited to produce such a study. This will be widely read from different angles, illustrating the intersectionality of numerous academic disciplines and practice-related fields in relation to disaster management. I have been very impressed with the engagement and remarkably activist approach taken by this increasingly prominent young author, whose affiliation with several foreign institutions seems to have further sparked the ardent desire to make a significant contribution to knowledge.

I commend and recommend this book to a wide readership, not just in Bangladesh, but worldwide, and congratulate the author for this timely intervention. As indicated, when it comes to the risks of disasters related to climate change, there is no longer time for endless ideologically focused discussions. This splendid study, combining theory and practice, is precisely what is needed at this moment to empower yet more young people to get involved in this enormous challenge to all of humanity, facing the brutal realities of climate change in a constructive, pointed manner.

Werner Menski
Emeritus Professor of South Asian
Laws, SOAS
University of London
London, UK
wm4@soas.ac.uk

Preface

According to the framework, good governance is essential to the efficient and long-lasting mitigation of catastrophe risks. Good governance requires both political will and robust institutions. Gains in development are at jeopardy from natural catastrophes, but human actions may also increase disaster danger. A proper institutional, policy, and legal structure is essential. By distributing funds, assuring and enforcing execution, assigning responsibility for failures, and making it easier for all essential stakeholders to become involved, good governance is anticipated to make disaster risk reduction a policy priority.

Bangladesh is one of the nations most vulnerable to natural and human-caused disasters, which can have devastating effects. In the field of disaster management, Bangladesh has shown itself to be an international leader. The government of Bangladesh (GoB) and the Ministry of Food and Disaster Management and Relief (MoFDMR) made national and international commitments to develop a National Plan for Disaster Management (NPDM) 2007–2015 and a Disaster Management Act to address disaster risks in a holistic manner. Furthermore, in 2008, the Standing Orders on Disaster (SOD) were updated with the stated goal of helping all relevant parties recognize their roles and responsibilities in disaster management.

Bangladesh should be commended for its proactive approach to disaster risk management and the incorporation of disaster risk reduction into development planning that has resulted from these efforts. However, in recent years, several initiatives have sought to "leapfrog" this barrier via the creation of more progressive plans of action and strategies that reflect contemporary ideas in disaster risk management and, to some extent, embrace the concepts of mainstreaming. There may be significant national and local progress toward mainstreaming if these plans and tactics are completely implemented.

Disaster management, as it is still usually referred as in department circulars and executive orders, is a legal obligation of individual line agencies and local governments. In reality, many UDMCs lack the resources to effectively reduce risks and incorporate them into their daily operations, and this is a problem for disaster management on the local level as a whole. From 2010–2015, the NPDM will work to include a "disaster preparation and management strategy" into the development planning process at all governmental levels. Nonetheless, this commendable objective is only established within the framework of the environment and natural resources sector, and the plan is even expressly complacent about catastrophe risk when discussing the particularly fragile agriculture sector.

Local government has struggled to successfully integrate disaster risk reduction into basic planning due to a number of planning process flaws, most notably the fact that development and investment plans are not always in place and, even if they are, are not always integrated. The government of Bangladesh does not include horizontal and vertical integration within the government into its overall planning process. Existing gaps in the planning system between different levels of government pose a potential impediment to mainstreaming, as they imply that locally identified needs are not necessarily reflected in higher-level plans and strategies, while there can be problems in implementing national policies and regulations at the local level.

The government of Bangladesh, as a signatory to the Hyogo Framework for Action (HFA) 2005–2015, is working hard to fulfill the HFA's objectives by promoting a culture that prioritizes catastrophe risk reduction rather than relief. In order to make disaster risk reduction (DRR) a fundamental concept in all key development sectors, political will and appropriate ability must be present. New laws have prepared the path for this shift. As a SAARC member state, Bangladesh has worked toward implementing the SAARC Comprehensive Framework on Disaster Management and Disaster Prevention by coordinating its mission with legislative approaches to disaster preparedness.

With new laws and reorganized institutions at every level of government, Bangladesh's political leadership has made DRR a priority for the first time in the country's history. The government's dedication to DRR reached its pinnacle in 2010, when it revised the SODs and approved the National Plan for Disaster Management. After that, in 2012, Bangladesh passed the National Disaster Management Act, which has had a profound impact on the country's preparedness for and response to natural disasters. The government, under Prime Minister Sheikh Hasina, has made significant strides toward its goal of mainstreaming disaster risk reduction by reducing the relief-centric strategy of the disaster management programme. Disasters in Bangladesh tend to bring out the donors. Government and donor agencies use the Local Consultative Group on Disaster and Emergency Response (LCG-DER) as their primary venue for making strategic decisions and exchanging ideas and information pertaining to disaster management. United Nations Development Programme, World Food Programme, and United Nations Children's Fund all work with local non-governmental organizations to carry out their aid programmes.

Bangladesh's disaster management system has been greatly improved thanks to the 2012 Disaster Management Act. The development process cannot be considered complete without including measures to reduce the danger of natural disasters. Bangladesh's disaster management community (consisting of government officials, local government committee embers, local representatives, local people, NGOs and others) has a firm grasp on the paradigm shift idea, and it has been included in the country's policy discourse. It was very difficult to coordinate the reaction to the tragedy since each humanitarian organization had to come up with their own plan, and the goals of those plans were too different to be coordinated. There were insufficient safe havens in case of floods or cyclones. In addition, the Department of Crisis Management has trouble getting enough trained staff mobilized in the event of a disaster. Furthermore, the infrastructure for data storage, compilation, and analysis was insufficient, and local disaster management committees lacked awareness and expertise concerning requirement assessment. Both the District and Upazila Disaster Management Committees lacked political leadership.

In order to avoid a relief-centric disaster management strategy, it seems that DRR and development in Bangladesh should proceed in tandem. To successfully move DRR into the realm of development, strong leadership and a shared vision of DRR as developmental rather than catastrophe management are essential. To reduce disaster risk and help vulnerable people manage their lives and livelihoods in a more secure and less risky way, DRR must be mainstreamed, meaning that all stakeholders, both state and non-state, must embrace DRR into their respective development plans. Financial institutions, international agencies, non-governmental organizations, and donor governments are all aware of the importance of focusing on disaster loss prevention programs, but they are unable to contribute to the government's direct funding facility for a coordinated disaster risk reduction programme. Donors, led by UNDP, work hand in hand with local NGOs to carry out each programmed initiative.

Savar, Dhaka, Bangladesh Mohammad Tarikul Islam

Acknowledgements

I must acknowledge my extensive debt to my teachers and colleagues in the Department of Government and Politics, Jahangirnagar University, including Prof. Md. Shamsul Alam and Prof. Bashir Ahmed, Prof. Joerg Friedrichs, Oxford Department of International Development, University of Oxford (my visiting research fellowship guide), Prof. Dr. Samita Sen, Faculty of History, at the University of Cambridge (my post-doctoral host), Prof. David Lewis, London School of Economics and Political Science, Dr. Michael Buehler, Department of Politics and International Studies, SOAS, University of London (Academic Host), Prof. Hafiz T. Khan, Visiting Professor of Oxford and Professor of Population Science at the University of West London, and my students particularly Zaki Faisal (UNDP Officer) for many enlightening discussions over the last couple of years. In all writings, I am benefited a lot from the critiques of academics, policymakers, and students, and this book is no exception. I should be thankful to my wife Tanzila Mustary (Masters in Political Science, University of Dhaka, and Principal Officer, Sonali Bank Limited), my son Tawseef Islam, and my daughter Toshana Islam for enabling me to concentrate on writing at home amidst the COVID pandemic.

I express adequately my gratitude to Ms. Nupoor Singh and Ms. Sharmila Mary Panner Selvam of Springer Nature for extending great cooperation.

Savar, Dhaka, Bangladesh Mohammad Tarikul Islam

About This Book

This book tries to unveil the nexus among disaster, governance, and development in Bangladesh and examines the legislative and institutional aspects in mainstreaming disaster risk reduction into the development planning. With the help of rich content analysis interpreting the disaster management history of the country, the book looks at the challenges associated with disaster management in the context of Bangladesh. The volume has discovered the most reasonable strategy on how to accelerate the paradigm shift from the relief culture to the DRR culture. It also takes a look at assessing the viewpoint on how political economy influences the government of Bangladesh on governance and institutional strengthening to help identify obstructions and opportunities for mainstreaming disaster management into development and how does governance work in the implementation of disaster management programmes in Bangladesh. The book also emphasizes on collaboration between public sector and private sector for the expansion of disaster risk reduction programme. It also shows how does the multi-level governance perceive as new addition to the approaches of governance for professionalizing disaster management. To what extent, policy framework has been developed in response to increased calls by both the people of the disaster-prone Bangladesh and the international community to recognize the disaster risk reduction as an institutional basis to administer the efforts of the government of Bangladesh in reducing the disaster risk for a safer community is narrated in the volume. This book will be useful for scholars, practitioners, and researchers of disaster management, environmental studies, development agencies, political science, public policy, development studies, governance, regional development, South Asian studies, and local government, particularly those interested in disaster, governance, and sustainable development.

Contents

About the Author

Dr. Mohammad Tarikul Islam is an Associate Professor of Government and Politics at the Jahangirnagar University in Bangladesh. Before joining the University in 2014, he worked with the UN Development Programme for seven years in various managerial capacities. Professor Islam acts as a Resource Person at Bangladesh Public Administration Training Centre. His edited book, *Human Security, Peace, and Development: South Asian Perspective*, which was released in India, has received a great deal of scholarly acclaim. His two significant textbooks, *COVID-19 in South Asia: Its Impact on Society, Economics and Politics* (Routledge) and *Local Government in Bangladesh: Contemporary Issues and Challenges* (Routledge), are forthcoming. Dr. Islam is a regular contributor to the South Asia Blog of the London School of Economics and Political Science (LSE), Oxford Political Review and ISAS, NUS. His articles regularly appear in some of the leading English dailies of South Asia such as the Daily Star, the Daily Independent, the Financial Express, the Business Standard, the South Asia Monitor, and the Khabarhub (Nepal). Professor Islam has held a number of prestigious positions (Postdoctoral Fellow, Visiting Researcher, Visiting Scholar) at the University of Oxford, the University of Cambridge and the School of Oriental and African Studies (SOAS) in the past.

Acronyms

ADPC	Asian Disaster Preparedness Centre
ADRC	Asian Disaster Reduction Centre
AFD	Armed Forces Division
APD	Academy for Planning and Development
BAF	Bangladesh Air Force
BARD	Bangladesh Academy for Rural Development
BBS	Bangladesh Bureau of Statistics
BCAS	Bangladesh Centre for Advanced Studies
BCSAA	Bangladesh Civil Service Administration Academy
BDRCS	Bangladesh Red Crescent Society
BFS&CD	Bangladesh Fire Service and Civil Defence
BGS	British Geological Survey
BIM	Bangladesh Institute of Management
BIRD	Bangladesh Institutes for Rural Development
BMD	Bangladesh Meteorological Department
BNBC	Bangladesh National Building Code
BPATC	Bangladesh Public Administration Training Centre
BS	Bangladesh Scouts
BUET	Bangladesh University of Engineering and Technology
BWDB	Bangladesh Water Development Board
CC	Climate Change
CCA	Climate Change Adaptation
CCC	Climate Change Cell
CCDMC	City Corporation Disaster Management Committee
CCDRCG	City Corporation Disaster Response Coordination Group
CDMP	Comprehensive Disaster Management Programme
CEGIS	Centre for Environmental and Geographical Information Services
CPP	Cyclone Preparedness Program
CPPIB	Cyclone Preparedness Program Implementation Board
CRA	Community Risk Assessment

CSDDWS	Committee for Speedy Dissemination of Disaster-Related Warning/Signals
DAE	Department of Agricultural Extension
DC	Deputy Commissioner
DDM	Department of Disaster Management
DDMC	District Disaster Management Committee
DDRCG	District Disaster Response Coordination Group
DER	Disaster Emergency Response
DFID	Department for International Development
DGHS	Directorate General of Health Services
DHTW	Deep Hand Tube Well
DIM	Direct Implementation Modality
DLS	Department of Livestock Services
DM	Disaster Management
DMA	Disaster Management Act
DMB	Disaster Management Bureau
DMC	Disaster Management Committee
DMIC	Disaster Management Information Centre
DMIN	Disaster Management Information Network
DMRD	Disaster Management and Relief Division
DoE	Department of Environment
DoF	Directorate of Forest
DoL	Department of Livestock
DoRR	Directorate of Relief and Rehabilitation
DOWA	Department of Women Affairs
DPHE	Department of Public Health Engineering
DRR	Disaster Risk Reduction
DRR	Directorate of Relief and Rehabilitation
DRRO	District Relief and Rehabilitation Officer
EOC	Emergency Operation Centre
EWS	Early Warning System
FAO	Food and Agriculture Organization
FFWC	Flood Forecasting and Warning Centre
FPMU	Food Planning and Monitoring Unit
GDP	Gross Domestic Product
GIS	Geographical Information System
GOB	Government of Bangladesh
GSB	Geological Survey of Bangladesh
GSB	Geological Survey of Bangladesh
HBRI	House Building Research Institute
HFA	Hyogo Framework for Action
ICT	Information Communication Technology
IDDR	International Day for Disaster Reduction
IMDMCC	Inter-Ministerial Disaster Management Coordination Committee

IMF	International Monetary Fund
INSARAG	International Search and Rescue Advisory Group
ISDR	International Strategy for Disaster Reduction
IVR	Interactive Voice Response
IWFM	Institute of Water and Flood Management
IWM	Institute of Water Modeling
IWM	Institute of Water Management
LCG	Local Consultative Groups
LDRCG	Local Disaster Response Coordination Group
LGD	Local Government Division
LGED	Local Government Engineering Department
LGRD	Local Government and Rural Development
MDGs	Millennium Development Goals
MoD	Ministry of Defence
MoFDM	Ministry of Food and Disaster Management
MoU	Memorandum of Understanding
MP	Member of Parliament
NDMC	National Disaster Management Council
NDRCC	National Disaster Response Coordination Centre
NGO	Non-Government Organization
NILG	National Institute for Local Government
NIPSOM	National Institute for Preventive Social Medicine
NPDM	National Plan of Disaster Management
PDMC	Pourashava Disaster Management Committee
PDRCG	Pourashava Disaster Response Coordination Group
PIO	Project Implementation Officer
PMO	Prime Minister's Office
PRA	Participatory Rural Appraisal
RDA	Rural Development Academy
RR	Risk Reduction
RRAP	Rural Risk Reduction Plan
RRAP	Risk Reduction Action Plan
SAARC	South Asian Association for Regional Cooperation
SDMC	SAARC Disaster Management Centre
SFA	SAARC Framework for Action
SOD	Standing Orders on Disaster
SPARRSO	Space Research and Remote Sensing Organization
UDMC	Union Disaster Management Committee
UDRCG	Upazila Disaster Response Coordination Group
UNDP	United Nations Development Programme
UNFCCC	United Nations Framework Convention on Climate Change
UNISDR	United Nations International Strategy for Disaster Reduction
UNO	Upazila Nirbahi Officer
UZDMC	Upazila Disaster Management Committee

UZDMP	Upazila Disaster Management Plan
WAPDA	Water and Power Development Authority
WDMC	Ward Level Disaster Management Committee

Chapter 1
Introduction

Natural disasters put development benefits in jeopardy, yet development decisions can raise the likelihood of a disaster. Because of this, every facet of development should work to lower rather than raise the danger of disasters. In this situation, governance is becoming more and more significant. Governance, which can be good or terrible, is one of the most significant elements impacting disaster risk. It is necessary to have adequate institutional, policy, and legal frameworks. Disaster risk reduction is required to be elevated as a policy priority, receive the necessary funding, be ensured and enforced, have responsibility for failures assigned to it, and be made accessible to all key stakeholders.

The word "DRR" refers to the process of preventing or lessening the effects of catastrophes. The United Nations, which has created a global plan to promote catastrophe risk reduction since it has been proven to be extremely cost-effective, has endorsed the project. "The conceptual framework of elements considered with the possibilities to minimize vulnerabilities and disaster risks throughout a society, to avoid (prevention) or to limit (mitigation and preparedness) the adverse impacts of hazards, within the broad context of sustainable development," is how the United Nations International Strategy for Disaster Reduction (UNISDR) is defined [1].

With contributions from leading international specialists in the subject, Culture and Disasters uses an interdisciplinary approach to investigate the cultural aspect of disaster. Since culture may be understood from many different angles, it is possible to critically evaluate the cultural boundaries of research as well as the challenges of merging many interpretations into DRR. Related topics including adaptation, coping, intervention, knowledge, and power relations cannot be properly understood without taking culture into account.

With serious effects linked with many natural and man-made dangers, Bangladesh is one of the most disaster-prone nations in the world. Bangladesh is extremely vulnerable to natural disasters due to its geophysical location, land characteristics, river system, and monsoon climate. Bangladesh's coastal morphology affects how risks affect the region. Natural disasters make coastal residents more vulnerable and impede their social and economic growth, particularly in the southeastern region.

M. T. Islam, *Disaster, Governance and Development*, Disaster Studies and Management,
https://doi.org/10.1007/978-981-99-1631-3_1

knowledge of the types of soil, plants, and seeds that are resilient to drought, floods, and cyclones. To lessen the likelihood of a poor harvest, farmers in marginal lands employ mixed cropping and intercropping practices. In several locations, houses are also constructed on elevated platforms to protect them above flood levels.

To cope with the disasters, communities have devised their own indigenous economic solutions. Typically, vulnerable households work to stockpile food, money, and other supplies that they can utilize when times are tough. Natural or wild foods from the forests, such as roots and berries, are used when food is scarce. Cultural difficulties include religious traditions and beliefs, which aid the community in recognizing disaster warning signals and serve as a conduit for passing down information and experiences from one generation to the next. This is transmitted through stories, legends, and other customs.

People may have a particular mindset or set of views based on their ideologies about right and wrong that have been refined by culture. The central idea of this significant work, disaster risk reduction (DRR), which denotes understanding that while natural disasters are inevitable, concerted action may, and in fact, must be taken to minimize risks, reflects this in a very obvious way. It goes without saying that numerous parties are involved, and for optimal effectiveness, they must learn to collaborate, frequently in novel ways. The National Plan for Disaster Management of 2007–2015, the Disaster Management Act of 2012, and down to the local level all contribute to the DRR culture of alertness that this study seeks to create.

The paradigm shift from the former post-disaster relief culture to the currently dominant DRR culture appears to be and is highlighted as being of utmost importance in this case. This entails taking into account both the advantages and disadvantages of integrating disaster management practices into integrated development structures through responsible forms of governance at all levels, starting with individual consciousness and self-help and moving up through various forms of collective and community-based processes and several layers of government structure. Political leadership is crucial in this situation as well as visionary leadership.

Through the persistent efforts of the government, local communities, development partners, and other stakeholders at all levels, Bangladesh maintains a culture of disaster resilience that keeps getting stronger. By (a) bringing about a paradigm shift in disaster management from conventional response and relief practice to a more comprehensive risk reduction culture and (b) strengthening the capacity of the Bangladesh disaster management system in improving the response and recovery management, the government of Bangladesh is making every effort to reduce the vulnerability of the poor to the effects of natural, environmental, and human-induced hazards to a manageable and acceptable humanitarian level. Below, a paradigm change in disaster management is explained in relation to Bangladesh.

The main goal of the Bangladeshi government's disaster management strategy is to improve the nation's ability to control risks associated with disasters while also coordinating emergency responses and recovery activities. It is performed through a thorough strategy that takes into account all dangers, all stages of disaster management, and all important parties. Knowledge building and policy support with and via

typically remarkably similar to current radio forecasting. There are few differences in the conventional knowledge system between different social classes.

Traditional community-level coping mechanisms include, among other things, maintaining a reasonable stock of dry food for survival in the wake of a disaster, keeping a certain amount of cash, stocking rice and vegetable seeds, moving livestock to relatives who live in less vulnerable areas or selling livestock based on indigenous predictions of flooding and other disaster events, agricultural production avoiding climatic disasters, and agricultural production avoiding climatic disasters. Social capital, which generally consists of kinship networks, mutual helps, and self-help groups, is one of the indigenous coping mechanisms. Many groups have swapped food and supplies during the crisis. Following the disaster, it is also possible to observe how the populace worked together to rebuild or restore the devastated infrastructures.

Additionally, there are numerous indigenous knowledges that can be used to lessen the effects of risks. According to the many academic works, locals are aware of how the climate is changing and how frequently disasters occur, and they create their own adaptation strategies based on their cultural practices and prior experiences. Using the knowledge passed down from their ancestors, people in Bangladesh constantly develop methods to coexist with the environment. Typically, this knowledge takes the form of social customs and actions. This information is expressed in many nations as songs, folktales, and proverbs, which typically become ingrained in cultural behaviors and beliefs.

Indigenous knowledge has its roots in the communities, yet throughout time, it is frequently impacted by other sources. Indigenous knowledge is described by the United Nations Ecosystem Program (UNEP) [4] as the knowledge that an indigenous community has amassed over generations of habitation in a particular environment. It refers to all types of knowledge that help a community maintain a stable standard of living. This can take the form of techniques, knowledge, customs, values, and cultural behaviors. Local and indigenous knowledge, as we have seen in the example of rural Bangladesh, has a direct connection to the natural world and the particular environmental situation. The value of indigenous customs is primarily supported by four arguments, according to the UNISDR [5] research.

(1) Knowledge regarding strategies to cope against natural disaster can be transferred from one community to another community in the similar situations.
(2) Affected community gets encouraged if there is an incorporation of indigenous knowledge in existing practices and policies. This empowers the members of the community to take the leading role in all disaster risk reduction activities.
(3) Information contained in indigenous knowledge can help improve project implementation by providing necessary information about the local context.
(4) Non-formal means by which indigenous knowledge is disseminated provides a successful model for the other education on disaster risk reduction.

It is widespread throughout practically all of the world. The influence of technology is felt in almost every facet of life. The community is regarded to benefit more from traditional practices than from scientific ones. The majority of populations in Bangladesh's disaster-prone regions, as well as in some circumstances, have deep

The nation has experienced nearly 200 disaster events since gaining independence in 1971, including cyclones, storms, water surges, floods, tornadoes, earthquakes, droughts, and other calamities that resulted in more than 500,000 fatalities and negatively impacted quality of life, livelihoods, and the economy [2].

The UNISDR [3] claims that Bangladesh leads the way among least-developed nations in giving disaster risk mitigation top priority in national fiscal planning. The Bangladeshi government has received praise for establishing a Comprehensive Disaster Management Program that views catastrophe risk reduction as an essential component of national fiscal planning, according to the aforementioned article. Bangladesh has adopted a general risk reduction model that encouraged national stakeholders to take into account both the risks of current disasters and the risks of projected climate extremes when constructing national and community resilience, with the help of the Comprehensive Disaster Management Programme. Disaster risk reduction model adopted by the government of Bangladesh has been elaborated below:

Disaster risk reduction model in the context of Bangladesh includes (a) assessing underlying risk; (b) initiating risk mitigation strategy; and (c) preventing the risk. Objectives of the disaster risk reduction model in Bangladesh are:

- Reducing existing risks (vulnerability and hazards).
- Adapting to changing risk factors (e.g., climate change).
- Preventing the further growing of risks through risk-conscious development.

It applies an integrated approach that:

- Considers the risk concept.
- Follows the principles of sustainability.
- Observes a multi-stakeholder approach.
- Considers DRR as a development issue.

Risk analysis and risk appraisal are the foundation for risk reduction measures. By examining possible risks, current vulnerable situations (physical, social, economic, and environmental), and accessible coping mechanisms, the risk analysis establishes the type and scope of risk. It produces a risk profile for the country, region, or site under consideration. The perception of hazards, as well as the weighting of catastrophe risks relative to other risks like health or economic concerns, are all included in the risk appraisal. Multi-stakeholder approaches have been a key component of the strategy, with a focus on Bangladeshi communities' efforts to reduce catastrophe risk.

Bangladesh's people, who have shown incredible fortitude in the face of natural disasters, should be celebrated. They have developed their own coping mechanisms to deal with the shock of disasters and have learned to live with hazards. In Bangladesh, the community has used coping mechanisms that we may consider commonplace over the ages with commitment and experimentation. If I use a flood as an example, you will see that the elderly in the village have a system of traditional knowledge that allows them to predict weather patterns in advance, identify the type of flood that will occur, and take precautions to lessen the risk of flooding. Their forecasts are

Chapter 1
Introduction

Natural disasters put development benefits in jeopardy, yet development decisions can raise the likelihood of a disaster. Because of this, every facet of development should work to lower rather than raise the danger of disasters. In this situation, governance is becoming more and more significant. Governance, which can be good or terrible, is one of the most significant elements impacting disaster risk. It is necessary to have adequate institutional, policy, and legal frameworks. Disaster risk reduction is required to be elevated as a policy priority, receive the necessary funding, be ensured and enforced, have responsibility for failures assigned to it, and be made accessible to all key stakeholders.

The word "DRR" refers to the process of preventing or lessening the effects of catastrophes. The United Nations, which has created a global plan to promote catastrophe risk reduction since it has been proven to be extremely cost-effective, has endorsed the project. "The conceptual framework of elements considered with the possibilities to minimize vulnerabilities and disaster risks throughout a society, to avoid (prevention) or to limit (mitigation and preparedness) the adverse impacts of hazards, within the broad context of sustainable development," is how the United Nations International Strategy for Disaster Reduction (UNISDR) is defined [1].

With contributions from leading international specialists in the subject, Culture and Disasters uses an interdisciplinary approach to investigate the cultural aspect of disaster. Since culture may be understood from many different angles, it is possible to critically evaluate the cultural boundaries of research as well as the challenges of merging many interpretations into DRR. Related topics including adaptation, coping, intervention, knowledge, and power relations cannot be properly understood without taking culture into account.

With serious effects linked with many natural and man-made dangers, Bangladesh is one of the most disaster-prone nations in the world. Bangladesh is extremely vulnerable to natural disasters due to its geophysical location, land characteristics, river system, and monsoon climate. Bangladesh's coastal morphology affects how risks affect the region. Natural disasters make coastal residents more vulnerable and impede their social and economic growth, particularly in the southeastern region.

M. T. Islam, *Disaster, Governance and Development*, Disaster Studies and Management,
https://doi.org/10.1007/978-981-99-1631-3_1

the government, as well as community-level interventions to target the most susceptible segment of the population, are two fundamental tenets of the government of Bangladesh's disaster risk reduction culture. The history of Bangladesh's disaster management makes it clear that, prior to 1996, the entire government was totally dependent on a relief-centric approach to disaster management, where assistance in the form of money and other goods was administered by the government authority.

In 1996, the political administration at the time first introduced the basic concept of risk mitigation. The political administration highlighted the necessity of lowering the danger of impending calamities in Bangladesh. The most frequently stated sign of the paradigm shift's success is a decrease in disaster-related death. The government has undertaken a number of initiatives to upgrade early warning signals for dissemination among those who are susceptible to catastrophe susceptibility in an effort to reduce the loss of lives and property. The institutional and legislative frameworks have been revised to broaden the notion of catastrophe risk mitigation. The National Plan for Disaster Management, the Disaster Management Act, and the Standing Order on Disasters are a few examples of legislation that has been developed in Bangladesh that pertain to disaster management.

Therefore, poverty alleviation initiatives should increase the vulnerable population's capacity to withstand the effects of catastrophes. Disaster risk reduction initiatives can also aid in reducing poverty and advancing development. So, it makes sense to incorporate catastrophe risk reduction into development and poverty alleviation strategies. Thus, reducing disaster risk is a challenge for Bangladesh's growth.

Understanding the Nexus Between Disaster Governance and Development from the Theoretical Viewpoint

Based on the opinions of experts, this section defines the connection between disaster, governance, and development. The following four ideas can be used to sum up how disaster and development interact: Development can: (1) make people more vulnerable; (2) make people less vulnerable; (3) interrupt development; and (4) make people more open to development.

When it comes to agricultural producers and their communities, [6]'s Adaptive Governance of Disaster offers a comparative analysis of policy instruments intended to address climate change, drought, and floods in relation to four case study regions: Alberta and Saskatchewan, Canada, Coquimbo, Chile; and Mendoza, Argentina. Tools for enhancing the livelihood capitals of agricultural producers are identified, their efficacy and adaptive governance evaluated, and suggestions are offered to enhance the suite of policy instruments [6].

The 2016 book Disaster Governance in Urbanizing Asia, edited by Michelle Ann Miller and Mike Douglass, examines the threat and effects of environmental disasters on Asia's urban inhabitants from the perspective of governance. It adopts a multi-sectoral, multi-disciplinary approach to disaster governance and places a focus on the significance of various stakeholders in disaster prevention, response, and recovery as well as the cascading consequences of disasters on Asia's cities. The contributors to this volume take a wide perspective on the many different causes of environmental disasters in urbanizing Asia, as well as the linked dangers and vulnerabilities. The book serves as an invitation to improve scholarship in the search for more efficient, all-encompassing, and inclusive goals for catastrophe preparedness, recovery, and development [7].

The 2015 book Cultures and Disasters: Understanding Cultural Framings in Disaster Risk Reduction, edited by Fred Krüger, Greg Bankopp, Terry Cannon, Benedikt Orlowski, and E. Lisa F. Schipper and featuring contributions from eminent experts from around the world, takes an interdisciplinary approach to disasters' cultural impact. To comprehend the significance of culture in risks and disasters, Sect. 1.1 examines theoretical and practical issues. Since culture may be broadly construed in a variety of ways, it enables us to critically explore the cultural constraints and the challenges of incorporating different interpretations into disaster risk reduction (DRR). Culture-related adaptation, management, intervention, knowledge, and power relations cannot be fully grasped when it is absent. The topics of culture resilience, how humans have cultivated culture in daily life, what makes a resilient culture, and how culture affects societal decision-making are all covered in Sect. 1.2. While it is natural for people to turn to tried-and-true techniques of disaster mitigation, this book explores the ongoing evolution of culture and ideas. Therefore, it raises the challenging question of how these coping mechanisms might be added to the DRR. The successful contribution of culture to catastrophe risk reduction and management within major scientific frameworks is reviewed in Sect. 1.3. The DRR includes several rules that disrupt the established theoretical and empirical frameworks by taking into account cultural will [8].

Disaster Risk Reduction Approaches in Bangladesh, an edited book by Shaw et al. [9], offers disaster risk reduction approaches in Bangladesh that demonstrate a methodical examination of disaster risk reduction in Bangladesh [9]. They include geological and hydro-meteorological hazards and cover all forms of disasters that can occur in Bangladesh. This book describes Bangladesh's disaster risk reduction (DRR) strategies using examples and lessons from important national and local initiatives, programs, and experiences. The content is based on a selection of documents that are readily available, a consultative workshop with academicians from various universities in Bangladesh that offer higher education programs in DRR (Dhaka University, BUET, Jahangirnagar University, Sylhet Shahjalal Science and Technology University and Independent University), as well as the editors' own knowledge and experience in the area. An academic analysis of field experiences is given particular priority, along with identifying important concerns and the policy importance of catastrophe risk reduction.

The authors describe Bangladesh's disaster risk reduction (DRR) strategies using examples and lessons from relevant national and local initiatives, projects, and experiences. An academic analysis of field experiences is given particular priority, along with identifying important concerns and the policy importance of catastrophe risk reduction. Students and scholars interested in environmental studies, disaster risk reduction, and climate change studies make up the book's main target audience. The book will provide readers a good notion of the direction that research is now taking in the area and will give them a foundational understanding of this significant subject in Bangladesh. Practitioners and decision-makers are another target group who can use collective knowledge in policy and decision-making. It is also revealed that due to service privatization, public support systems and services have substantially deteriorated, and foreign disaster relief and reconstruction monies have not kept up with the rising need. International humanitarian assistance for poor nations has increased (rather than decreased) their reliance on rich nations, posing significant risks for the recovery after earthquakes in India and Nepal. The nature of help, which has mostly concentrated on short-term, immediate relief initiatives, and a lack of political will to adopt the necessary norms and standards are likely to be the causes of dependence on outside assistance.

In their edited book Climate Change Adaptation Actions in Bangladesh, Shaw et al. [9] describe these efforts within the context of Bangladesh. Internationally renowned professors and researchers in the field collaborated to write the most sophisticated textbook on cutting-edge structural wind engineering which includes a wide variety of issues related to structural wind engineering. The authors describe the climate change adaptation (CCA) initiatives in Bangladesh using illustrations and takeaways from various initiatives and programs in the nation. The authors' own knowledge and experience in the field, a consultation workshop with academicians from several universities that offer higher education on catastrophe risk reduction and climate change adaptation, and a selection of publicly available materials all contributed to the content.

The management of local knowledge is another point of emphasis in this study on catastrophe management. Probably because it is the most evident and useful element of local knowledge, local disaster management knowledge was frequently primarily associated with local technical knowledge. Local building practices, the usage of particular building materials, and their synthesis are all covered by local, technical skills. In addition to local, technical knowledge, there are other categories of knowledge, such as agricultural and environmental knowledge, sociocultural knowledge, and historical knowledge. Because it is so deeply ingrained in people's way of life and worldviews, this local, non-structural knowledge is difficult for outsiders to recognize. The most extensively researched subject is ecological knowledge, which relates to local knowledge of natural resources and addresses the significance that scholars attach to sustainable development. Studies have emphasized the depth of regional environmental knowledge and demonstrated how indigenous practices like agroforestry and polyculture help preserve ecological variety. Despite its significance, sociocultural and historical knowledge is frequently disregarded in studies of natural disasters. Sociocultural knowledge comprises information about the social, political, economic, and spiritual facets of life as well as the sociocultural environment in its broadest meaning. Social relations in the area have a significant impact on how people perceive and react to natural risks. The way that people react to interventions depends on their views about the outside world (the regional, national, and international entities likely to intervene in catastrophe responses). This pertains to knowledge about development initiatives.

The efficiency of Union Disaster Management Committees for organizing and coordinating relief and recovery efforts centered on storm Aila. Uttar Bedkashi is one of the unions in Khulna district's Koyra Upazila that was impacted by Cyclone

Aila. Following Cyclone Aila, the Union Disaster Management Committee (UDMC) carried out tasks such as holding 18 meetings, awarding certificates to 37 NGOs for relief and rehabilitation efforts, conducting assessments on Aila five times, and providing assistance to the Bangladesh Army and the GoB in rebuilding damaged dams surrounding this Union.

The study also noted that the UDMC local-level organization was more or less effective, but after speaking with residents, it became clear that many of the locals are unaware of the institutional structure and how it operates. Locals, especially from vulnerable groups, claimed that they have very little access to the discussions and choices made at UDMC meetings. This basically indicates that the local vulnerable group members have little knowledge of the duties, responsibilities, and operations of the disaster management committee. There was no proof that UDMC had any involvement prior to the accident.

The cooperation of many state departments, the UN, NGOs, and even the private sector at the national level gives an apparent good indication of the governments' involvement in disaster management. However, the UDMC's response to calamities is not always effective. The local elites claim that this committee's inability to execute effective prevention initiatives or programs that would assist communities become less vulnerable to dangers in the future is caused by a lack of financial and material resources. It appears that there is still a lack of preventative measures and activities aimed at enhancing community resilience to natural disasters, particularly with relation to the execution of various interventions. The community-level consultation revealed a general opinion that disaster risk management is still a low priority and is not adequately integrated into the many programs being carried out by Union Parishad.

In his book At Risk: Natural Hazards, Vulnerability, and Disasters, Wisner (2004) discusses catastrophe as a warning of the failure of mainstream "progress," not as an anomaly. Two analytical models are offered as methods for comprehending vulnerability. One establishes a "progression of vulnerability" that connects far-off "root causes" to "unsafe conditions." The other attempts to explain why some households are more vulnerable than others by utilizing the ideas of "access" and "livelihood." Case studies related to drought, biological hazards, floods and landslides, cyclones, earthquakes, and volcanoes are presented in Part II. The book also offers guidance for healing and preventative measures to make the world safer. A set of recommendations is included at the end to help the national authority reevaluate development planning and programs in the context of DRR. The vulnerability to and risks from natural hazards have been increasing in developed and developing countries despite advances in knowledge and technology (such as satellite coverage or surveillance techniques), and this may be the case even though the frequency and magnitude of hazard events have remained constant. In other words, rather than the actual number of disasters caused by natural hazard occurrences, it is the effects of these events on people and property that have been rising. Changes in people's social, economic, cultural, political, and environmental settings have an impact on risks and susceptibility. For decades, both the frequency and cost of disasters have risen as a result of development efforts and a lack thereof. For instance, resource allocation and distribution among various groups of people as well as the emergence of new natural hazard hazards have been influenced by development processes (such as the construction of roads and dams). The most vulnerable among the displaced are compelled to reside on marginal land and in dangerous regions as a result of population and socioeconomic pressures.

In addition to serving as a resource for lessons on how to reduce risk and exposure to hazards and to meet the challenges of tomorrow, authors offer direction, policy orientation, and inspiration. It is aimed at those who are involved in and interested in catastrophe risk reduction and sustainable development. Anyone, anyone can be impacted by natural disasters. Due to their social, economic, and environmental fragility, people are at risk from hazards, which must be taken into account if sustainable development is to be realized. Therefore, everyone is concerned in disaster risk reduction, from farmers to leaders of state, from villages to attorneys, from farmers to foresters, and from meteorologists to media executives.

This understanding of humanitarian help and government assistance has caused a paradigm shift away from only emphasizing relief and toward disaster preparedness. This tendency raises a few problems, such as why modern technology and a wealth of data are often unable to reduce the risk posed by natural disasters. Is there a knowledge gap? Knowledge not put to use? Is knowledge applied improperly? How are the disaster victims, who are becoming more numerous, going to manage the shortage of public goods? Does sensible preparation for natural hazards necessarily result in sensible response? Whose interests are promoted by the study of natural disasters and hazards? Does this research serve the needs of those who are at danger or the interests of the collaborating technical and research organizations, which stand to gain from advancing science and technology?

UNISDR [1] emphasizes the regional strategy and results framework for comprehensive disaster management in its annual report living with Risk-A Global Review of Disaster Reduction Initiatives (CDM). Contributing to the Caribbean's sustainable development is the aim of CDM. Because of this, it is evident in the disaster management community that good risk management of hazards is a prerequisite for sustainable development in these Small Island Developing States (SIDS). As a result, there is an obvious connection between the SIDS Programme of Action (POA) agenda and the growing concern about natural hazard risk management around the world. The necessity for targeted interventions in disaster loss reduction capacity building is becoming more and more recognized on a regional political level. High-level decision-makers from the public and corporate sectors participated in national dialogues supported by the CDM promotional agenda over the past year in order to adopt the strategy and define national action plans to advance the process at the local level.

Disaster management stakeholders are also on board for the journey to incorporate disaster reduction principles in our development planning policies and programs, according to a UNISDR publication. Their intentional participation in the process of strategy design and revision has made this much simpler. The manuscript also discusses the support that the Caribbean Disaster Emergency Response Agency

Coordinating Unit (CDERA/CU) has requested from the United Nations Development Programme (UNDP) and the United States Agency for International Development (USAID) in order to reevaluate the national government's capacity to advance disaster risk reduction (DRR) throughout their development process. The majority of research on how people react to and adapt to natural disasters and hazards advanced more quickly in developing nations than in developed ones; much of it, particularly on drought, focused on indigenous peoples, small-scale farmers, and herders, and directly questioned the policies and practices of mainstream academia, the media, governments, and aid organizations. In the Himalayas, similar effort started in the early 1980s. Nevertheless, up until recently, local knowledge and practices were disregarded by mainstream Disaster Management Institutions and the literature on natural hazards and disasters. Local knowledge's existence and value were hardly discussed. Most academic research, both domestically and abroad, has focused on the most recent, "advanced" geophysical knowledge and technology systems as the most efficient disaster response techniques. The substantial body of research on local knowledge was still considered marginal, just like many of the populations whose knowledge it was. 1Part of this can be attributed to the vast gap between technical and social perspectives and the precedence given to the "expertise" approach, which places an emphasis on formal education and degrees rather than life experience. But since the 1980s, a growing number of organizations have realized how critical it is to incorporate "local knowledge" into growth. The same is true for disaster management, as more research projects, national and international NGOs, and UN agencies are starting to consider local knowledge and its stakeholders. Numerous NGOs have been founded locally, regionally, and internationally to deal with these problems or to take up advocacy on behalf of individuals who are vulnerable. It remains to be seen how much less subject to marginalization by national and international disaster management strategies the most recent work on local knowledge and related participatory disaster management approaches are than prior work.

It explains how DRR and development are related. In low- and middle-income nations, natural catastrophes—that is, disasters brought on by natural hazards such cyclones, droughts, floods, earthquakes, landslides, and volcanic eruptions—are frequent and pervasive. Both substantial human casualties and significant harm to communities, infrastructure, and economy at large can result from them. We must act to safeguard human life and lessen suffering out of ethical and humanitarian obligations. The author was accurate to note that numerous experts and assistance groups have identified natural disasters as a significant danger to sustainable development.

Model policies and programs for mitigation, hazard mapping, and vulnerability assessment are at an advanced state of development, according to the publication's results. The foundation of this change is horizontal cooperation, employing indigenous experiences and knowledge. We believe that the link between catastrophe reduction and sustainable development must be based on two fundamental principles: sustainability and self-reliance. Information, communication, and technology (ICT) offer a great platform for establishing the relationships with partners, exchange of information, and numerous accesses required to drive a sustainable development program with considerations for catastrophe reduction.

The Department of Disaster Management, a subsidiary department of the devoted Ministry of Disaster Management and Relief, is referred to as part of the institutional setup. The community-driven risk reduction initiative, in which the community works with government authorities to continue risk reduction measures, is the best aspect of the government of Bangladesh's ongoing paradigm change. Communities have been shown to be more resilient and to be incorporating disaster risk reduction and climate change adaptation methods. The concept of risk reduction and its advantages are gradually becoming understood by populations that are vulnerable to disasters. Communities at the municipal and governmental levels are outfitted with training, tools, and protocols on how to deal with these calamities when they occur. Over the years, coping mechanisms that utilized their ancestors' knowledge and methods have driven them.

Volunteers in Bangladesh's disaster-prone regions are all more knowledgeable on how to survive disasters with minimal damage and rebuild swiftly to reduce harmful effects. On top of that, we must state that while a culture of risk reduction is taking hold across the nation, the government of Bangladesh should embrace a complete strategy to disaster management that includes emergency response management that is centered on providing relief. The relief-centered strategy should be maintained in light of the fact that catastrophe victims do survive both during and after shocks from disasters. Here, the government of Bangladesh's instructions for reaction and recovery management includes the following values [10]:

1. Standardization of relief package.
2. Relief stocking with an exercise of needs mapping during non-disaster times.
3. Store house in each and every disaster-prone Upazila headquarter.
4. Coordination committee should be advised to remain vigilant during disaster seasons.
5. Standardization of recovery package.
6. Development of guideline for identifying disaster victims.
7. Mobilization of resources for recovery programme.
8. Preparation of monitoring and Evaluation guideline.

In its monthly journal NIRPADA Barta, NIRAPAD (1997), a disaster research network in Bangladesh, emphasizes Bangladesh's ongoing disaster preparedness efforts and disaster response management system. The article discusses how the Bangladeshi government collaborates with neighborhood NGOs to disseminate early warnings and handle coordinated disaster response. The paper investigated the lack of dynamic coordination between local administration and local government, finding that local administration did not seek input from local government when deciding on an action plan for managing disaster response. This article was written after a comparison of Bangladesh's preparedness and reaction efforts.

Mahbuba (1995) analyzes the existence of the native coping mechanisms of the rural population, particularly rural women, during floods in Bangladesh in her book Coping with Floods: The Experiences of Rural Women in Bangladesh. Flood-related catastrophe risk reduction may be the turning point for minimizing significant investment after disaster events. Bangladesh's experience handling floods throughout the years has been very unfortunate because the sole goal of the reaction and recovery has been to save lives by providing life-saving aid. Even the current government's BNP did not place a high premium on long-term recovery. This report also makes recommendations for implementing a flood preparedness program while taking into consideration rural women, who are among others the most disaster prone.

In his book Disaster Management Handbook for Bangladesh, Rahman (1993) outlines the many natural disasters and their effects as well as the disaster management initiatives implemented by the government and other interested parties. It describes the history of the nation's vulnerability to disasters in light of its geographical setting. The author also covers the application of the disaster preparedness program, community awareness levels, the government's legislative framework, and disaster management as a whole. A significant disparity between structural and non-structural arrangements in disaster response and recovery is also revealed in the book. Bangladesh is a highly lucky nation since its people possess local knowledge, bravery, and social capital. As a result, the danger of disaster has decreased, even if Bangladeshi society still debates the value of risk reduction. According to the participatory discourse, taking into account local knowledge in terms of practices and contexts can help implementing organizations better plan for and carry out disaster preparedness activities; it can also specifically help improve project performance and project acceptance, ownership, and sustainability. This means that, in the long run, from both a financial and a social perspective, recognizing, accounting for, and honoring local knowledge contribute to cost-effectiveness—especially in the context of complex, changing, and expanding dangers. First off, economies of scale are founded on the notion that various resources are found on different scales and that people perform better on some dimensions than on others from a financial perspective. Beyond the contradiction between municipal and state management levels, solutions in resource management, development, and disaster management must incorporate cross-scale institutional linkages. Understanding local customs and knowledge can assist determine what is required and acceptable locally and how to enlist people's participation to win their support for outside action. When it makes sense to do so, building on local knowledge and practices can reduce reliance on outside assistance. Locals maintain continuity and can keep an eye on the measures conducted.

In his book Disaster Preparedness Recommendations for Bangladesh, Cuny (1988) makes a number of suggestions for improving disaster preparedness in Bangladesh, placing particular emphasis on the roles and responsibilities currently held by the government of Bangladesh and other pertinent stakeholders in the field. In rural Bangladesh, he discusses the degree of community awareness and participation in disaster preparedness. The author's article was completed with statistical information about the community awareness program implemented by many stakeholders, including the government, and the percentage participation of the community.

References

1. UNISDR (2004) Terminology: basic terms of disaster risk reduction. UNISDR, Geneva
2. DDM (2012) Disaster management: response and recovery in 2011. Government of Bangladesh, Dhaka
3. UNISDR (1992) Global facility for disaster reduction and recovery: a partnership for mainstreaming disaster mitigation in poverty reduction strategies. UNISDR, Geneva
4. UNEP (2008) Disaster and environment- collective actions for future. UNEP HQ, Nairobi
5. UNISDR (2010) DRR and sustainable development: everyone business. UNISDR, Geneva
6. Hurlbert M (2018) Adaptive governance of disaster: drought and flood in rural areas. Springer International Publishing, Switzerland
7. Miller MA, Douglass M (2016) Disaster governance in urbanising Asia. Springer International Publishing, Switzerland
8. Krüger F, Bankoff G, Cannon T, Orlowski B, Schipper E, Lisa F (2015) Cultures and disasters: understanding cultural framings in disaster risk reduction. Routledge, UK
9. Shaw R, Mallick F, Islam A (2013) Disaster risk reduction approaches in Bangladesh. Springer, Japan, Tokyo
10. DMB (2009) Response and recovery guidelines. Disaster Management Bureau. Dhaka, Government of Bangladesh

Chapter 2
Concept of Disaster, Governance, and Development

2.1 Introduction

The term "disaster risk reduction" (DRR) is a methodical strategy for analyzing potential disasters and taking preventative measures to lessen their impact. Here, it has been heavily impacted by the plethora of published research on vulnerability that has amassed since the mid-1970s and tries to minimize socioeconomic vulnerabilities to catastrophe by addressing the environmental and other hazards that generate them. NGOs working on development and aid have an equal share of the responsibilities. It shouldn't be an afterthought or ad hoc addition to the way these groups function. DRR covers a lot of ground, going into far more detail than the average emergency management plan.

2.2 Development of Disaster Risk Reduction Concept

More research into the causes of catastrophes and the adoption of systemic, comprehensive strategies to mitigate their effects have been hallmarks of the disaster management field since the 1970s. The most recent development in this field is known as disaster risk reduction (DRR), and it has become the standard by which emergency preparedness is measured. While DRR is a newer concept formally, it draws on centuries-old ideas and methods. Numerous governmental and non-governmental groups have adopted it, including international organizations, states, and catastrophe planners.

Defining and explaining DRR in detail has been challenging because of the breadth of the concept, although the general notion is straightforward. While there will inevitably be disagreements over precise terminology in the academic literature, "hazard reduction" is often understood to refer to the systematic implementation of policies, methods, and practices designed to lessen societal exposure to hazards. The phrase "disaster risk management" (DRM) is commonly used interchangeably

M. T. Islam, *Disaster, Governance and Development*, Disaster Studies and Management,
https://doi.org/10.1007/978-981-99-1631-3_2

with "risk management" (RM) to refer to the same thing: a methodical strategy for detecting, analyzing, and mitigating the many risks that might arise from natural and man-made disasters. It is more suited to the real execution of DRR initiatives and the operational components of DRR [1].

There has been a rise in the number of voices demanding more precise definitions of DRR's constituent parts and of measures measuring progress toward resilience. Days after the 2004 Indian Ocean earthquake, the United Nations held the World Conference on Disaster Reduction (WCDR) in Kobe, Japan. There, member states and international development partners embraced DRR as a key force to lower underlying risk. The World Conference on Disaster Reduction kicked off efforts to get international organizations and national governments to reassess their conventional disaster management strategy in light of the principle of risk reduction. The World Conference on Disaster Reduction (WCDR) ratified the Hyogo Framework for Action (2005–2015) as the first stage in this process (HFA). This DRR framework is the first of its kind to receive universal approval [2].

HFA lays out a sequential set of goals (result–strategic goals–priorities), with five priorities for action that aim to "catch" the primary areas of DRR intervention. The United Nations and its member states can assess their progress toward the Hyogo Framework for Disaster Risk Reduction at each biannual Global Platform for Disaster Risk Reduction. Its inaugural meeting was place in Geneva, Switzerland, from June 5 to 7, 2007. The United Nations (UN) declared the 1990s the International Decade for Natural Disaster Reduction, which generated a number of efforts that helped improve and promote the concept on a global scale [3].

The purpose of disaster management is to lessen or prevent losses, help those affected by disasters as soon as possible, and get things back on track as soon as possible. The disaster management cycle demonstrates the ongoing efforts of governments, businesses, and civil society to prepare for, mitigate the effects of, and recover from disasters. Taking the right steps at any stage of the cycle can improve your chances of being ready for the next stage, receiving early warnings, being less at risk, or preventing a disaster altogether.

It is a common complaint in DRR's academic literature that cultural factors are overlooked during the strategy's development and implementation stages. The cultural significance of disaster preparedness was especially evident after the 2004 Indian Ocean Tsunami. In 2004, when the tsunami slammed the coasts of south Asian countries, certain communities with indigenous knowledge surrounding tsunami were successfully preserved, whereas migrants and tourists who did not have local knowledge were greatly harmed. Uneven development, structural disparities, and inadequate disaster preparedness systems and infrastructure contribute to varying degrees of vulnerability when natural or technological hazards interact with human populations. Misunderstandings between institutions and local populations can be costly and detrimental to disaster risk reduction and recovery efforts if planning and recovery are carried out without taking local viewpoints or cultural worldviews into account. At the same time, Bangladesh has been held up as an example of how to handle such problems elsewhere in the world. Because of this, the word "disaster management" has not only grown in prominence and sophistication across a range of academic fields but also in its most fundamental terms.

It's possible that a person's ideology, or their sense of what constitutes good and evil, has been shaped by their culture. This is mirrored in the book's central idea, disaster risk reduction (DRR), which acknowledges the inevitability of natural catastrophes while acknowledging the potential for and necessity of collaborative effort to mitigate their effects. Naturally, this involves a wide variety of parties that will need to figure out how to collaborate, sometimes in novel ways, in order to maximize effectiveness. The DRR culture of vigilance that this research supports permeates all tiers of government, from the national (through the National Plan for Disaster Management, 2007–2015) to the state (via the Disaster Management Act, 2012) to the municipal (via the DRR Act, 2014).

The shift in emphasis from disaster relief culture to DRR culture appears to be the most significant topic covered here. This necessitates thinking about the positives and negatives of incorporating disaster management processes into integrated development structures via responsible forms of governance at all levels, from personal awareness and self-help to various forms of collective and community-based processes and several tiers of government structure, all the way to the top. Political leadership as visionary management is also crucial in this regard.

There will always be opportunities to interfere with, exploit, or otherwise thwart the complex processes that must be in place. The success or failure of activist actions depends on constant attention to developing and testing novel approaches. The problems are insurmountable, but tying the entire area to the development processes is more proof that climate change contributes to novel facets of human progress in a wide range of fields. There will be clear synergies between disciplines as a result. An all-encompassing examination will require the contributions of nature scientists, the participation of human scientists, and a variety of governance systems, as well as the results of international solidarity initiatives. Bangladesh should put all of its abundant supplies to good use.

In order to lessen people's risks, the study stresses the need of concentrating on how people's livelihoods might be made more stable, safer, and, if required, replaced. People's perceptions of health risks are influenced by local traditions, beliefs, and social practices, which may or may not correspond to the expectations of public health interventions. This calls for a balancing act between local health beliefs or everyday practices and public health interventions. More and more people, organizations, and governments need to work together to ensure smooth operations during times of crisis. Organizational culture has a significant role in disaster risk reduction because it affects how various players perceive a crisis.

The process of finding reliable and understandable emergency information may be impacted by the cultural norms of the organization's many players. Comprehension of the variations between cultural organizations and the understanding of the interaction networks by agents is crucial for developing successful emergency management solutions. Culture in the workplace is increasingly being recognized as a major determinant of an organization's success. Culture in the workplace affects how emergency responders see the issue, prioritize their work, and communicate with one another.

To reduce the likelihood of catastrophes or lessen the impact they have on individuals, communities, and infrastructure, public policy and planning are integral parts of the disaster management life cycle. The disaster management improvement process begins with the hazard identification and risk assessment phase and continues with the mitigation and readiness phases as the catastrophe approaches. A community's preparedness and mitigation efforts might benefit greatly from developmental considerations. Culture was clearly a factor for the survival of communities in some cases, and a barrier for effective DRR initiatives, in others. It follows that cultural factors can either heighten or lessen a population's susceptibility to catastrophes. Moreover, the aforementioned case studies demonstrated how insufficient attention to cultural factors in the affected community might impede successful DRR tactics, hence increasing rather than decreasing the vulnerability of the affected population.

2.3 DRR and Public Policy: Interplay

Within the broader context of sustainable development, disaster risk reduction is a conceptual framework that aims to reduce societal vulnerability by systematically avoiding (prevention) and minimizing (preparedness and mitigation) the negative effects of hazards. DRR expands the scope of emergency preparedness to include assessing risks, creating and sharing new information, bolstering public support, establishing enabling institutional structures, optimizing resource management, and enhancing the speed and effectiveness of alerts and responses. Over time, the method has shifted from reactively "managing" disasters to taking a more preventative and systematic approach to lowering risk.

Humanitarian aid groups often deal with immediate needs once a disaster strikes, whereas development agencies focus on longer-term projects, with the understanding that such catastrophes are anomalies. Much unfinished business exists in the realms of preparedness, prevention, and resilience. It is not uncommon for initiatives to prioritize strengthening the "strengths" and "opportunities" rather than reducing the "risks". More and more lives and property are being lost every year because of severe events, but thankfully, international organizations and national governments are beginning to see the value of incorporating DRR into their development initiatives. DRR unites the assistance and development sectors by bridging the gap between them. The following are some of the many facets of catastrophe risk reduction:

Disaster mitigation: Measures were taken, both formally and informally, to mitigate the effects of natural disasters; for instance, planting mangroves to lessen the threat of tidal surges or educating students about the dangers of climate change.

Early warning: The dissemination of accurate and timely data that empowers individuals to lessen the severity of potential risks. Multi-hazard early warning systems are the norm, and they necessitate the actual ownership and engagement of communities and other stakeholders, such as giving residents advance notice of a typhoon or tropical storm.

Disaster preparedness: Community action teams backed by National Society contingency planning and regional and/or international reaction teams are examples of measures that help ensure a prompt and effective "first line" of response supported by volunteers, branches, regional and national capacity.

The word ***"recovery"*** is used to describe the post-disaster decisions and measures done to return the afflicted community to its predisaster state of living circumstances, if possible, while also facilitating the required improvements to decrease catastrophe risk.

Support to livelihoods: Improvements in nutrition and stockpiles in the face of drought are possible thanks to projects that fortify or diversify livelihoods, therefore empowering people and families to use risk mitigation techniques.

2.4 Phase-Wise Response Framework of Comprehensive Disaster Management

The response framework needs to be discussed in light of the three stages of disaster management as a whole. Our findings from this phase-by-stage model for catastrophe response are discussed (Fig. 2.1).

With the help of the community, the predisaster phase of the response framework created by Trobe and Davis entails identifying the probable risk associated with future catastrophes and developing an action plan to minimize that risk. Rules and regulations, as well as collaboration with relevant government agencies and non-state actors, are given high priority during the predisaster period. The following steps are recommended at this stage: (a) a community risk assessment; (b) a risk mitigation plan; (c) the modernization of early warning; (d) the capacity building of volunteers for search-and-rescue operations; (e) the preparation of emergency and early recovery guidelines; (f) the construction of community shelter centers; (g) the stockpiling of emergency relief items at the Upazila headquarters of disaster-prone areas; (h) the raising of awareness about how to cope with the challenges of Phase-II of the response framework describes in-depth methods for handling emergencies. Disseminating early warning signals, conducting search-and-rescue operations, and distributing relief packages are all common facets of emergency response management. In this step, we conduct a needs assessment to establish the scope of the disaster's impact and determine which victims will receive further support. It is crucial at this stage for the national government to act as a coordinator for the many relief organizations so that no one organization ends up offering the same type of aid again.

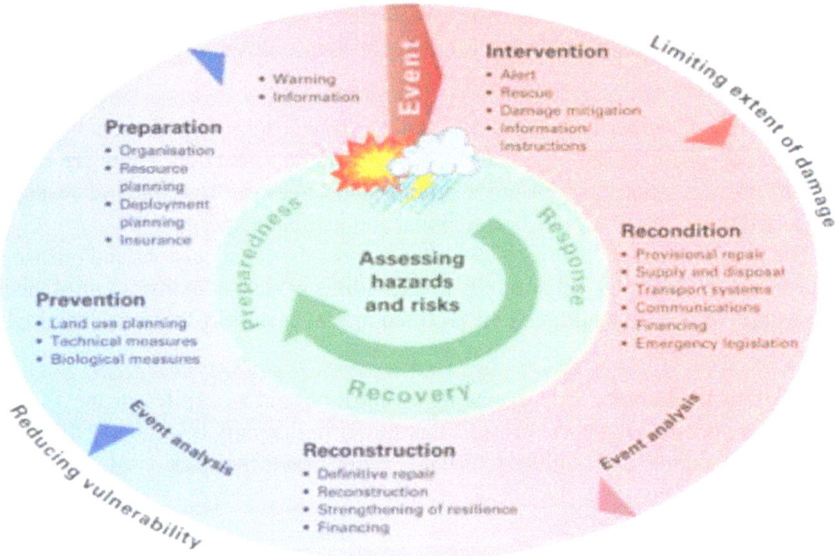

Fig. 2.1 Phase-wise response framework of comprehensive disaster management [4]

Phase-III of the phase-by-phase catastrophe response paradigm concerns the potential for long-term recovery. At this point, the government must conduct a thorough needs assessment across all sectors (including agriculture, housing, livelihood, environment, and so on). A long-term, sustainable recovery program will be initiated at this stage, at the victims' request. During this time, we'll do things like (a) enforcing recovery guidelines, (b) activating a disaster management coordination mechanism, (c) standardizing a recovery package based on sustainability principles, (d) mobilizing resources, (e) implementing a recovery strategy with input from affected communities, (f) monitoring and evaluating progress, and (g) conducting a social audit.

2.5 Disaster Risk Reduction and Global Policy Enforcement

There are five main obstacles to systematic action on DRR, as identified in the Hyogo Framework of Action (2005–2015), which was adopted by 168 countries at the UN World Conference on Disaster Reduction: (a) governance; (b) risk identification, assessment, monitoring, and early warning; (c) knowledge management and education; (d) reducing underlying risk factors; and (e) preparedness for effective response and recovery. At the World Conference on Disaster Reduction held in Kobe, Hyogo, Japan, in January 2005, a 10-year plan to make the world safer from natural disasters

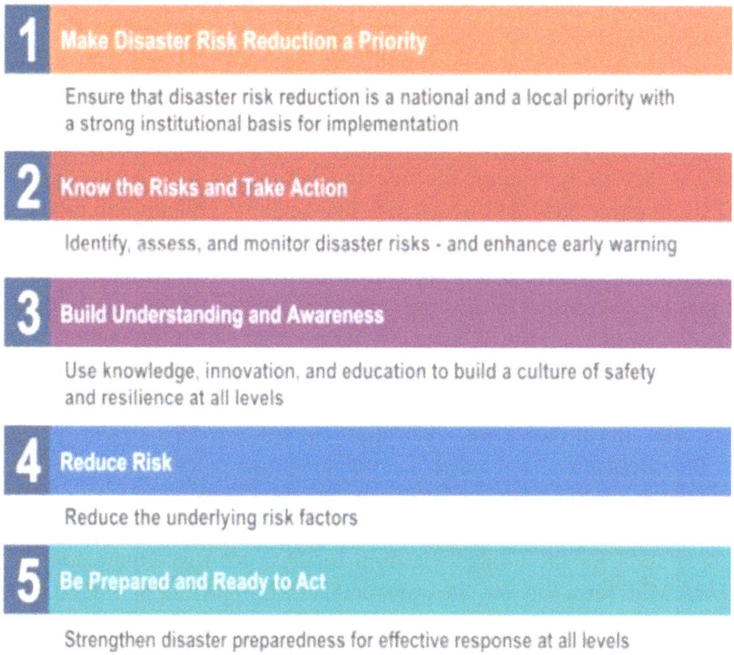

Fig. 2.2 HFA five priority of action [5]

was approved. This plan is known as the Hyogo Framework for Action (2005–2015). The Hyogo Framework serves as a worldwide guide for catastrophe prevention initiatives throughout the next decade. All signatory countries are required to report on their progress toward the five areas of focus listed below at regular intervals (Fig. 2.2).

In order to better prepare for and respond to catastrophes, research is viewed as an equal partner in HFA's five priorities of action for disaster risk reduction. As an illustration, the HFA's second priority of action, "learning about the unfavorable effect of catastrophe and learning suitable measures to manage such risks," captures the core of research and innovation. Adopting laws and regulations to include DRR into the government's development process is intended to simplify the field.

The same is made clear by HFA's top-tier, immediately applicable course of action. DRR is a process that aims to reduce the likelihood of devastating outcomes by preparing resilient communities to implement DRR measures. The national government that signed the HFA shall operate in good faith, keeping a close eye on developments and being ready to take any required action in light of the five HFA action plans.

In 2015, when the NPDM 2010–2015 was winding down, the "Sendai Framework for Disaster Risk Reduction" (SFDRR) was adopted at the 3rd United Nations World Conference on Disaster Risk Reduction in Japan. Since Bangladesh is an SFDRR signatory, the NPDM 2016–2020 reflects the country's commitment to adapting SFDRR's goals to domestic concerns. The "Asian Regional Plan for Disaster Risk

Reduction" (ARPDRR) was approved during the Asian Ministerial Conference on Disaster Risk Reduction in New Delhi in November 2016, and it is consistent with SFDRR. Thus, the NPDM 2016–2020 varies from the NPDM 2010–2015 in that its structure is consistent with newly created global and regional frameworks. All UN member states are using the Sendai Framework for Disaster Risk Reduction (SFDRR) as their primary guide for modern DRR [6].

SFDRR recommends that disaster risk management policies and procedures be founded on a thorough comprehension of disaster risk in all its forms. This information may be put to use in a variety of predisaster contexts, including risk assessment, prevention, mitigation, and the design and implementation of adequate preparedness and effective response. Four theme categories under which the country's primary strategic initiatives have been outlined to satisfy the first priority area of the SFDRR, "Understanding Disaster Risk," which is the foundation for all measures taken to withstand risks. In particular, we will be discussing: 1. Information and data management in the aftermath of a disaster, 2. Assessment of potential risks, weaknesses, and capabilities, 3. Last but not least, number four is dedicated to research and knowledge-creation initiatives. Mapping and/or zoning.

In preparation for the Global Platform for Disaster Risk Reduction 2022 and the SFDRR Mid-Term Review, the government of Bangladesh has asked all nations to reconsider the original Vision and Policy of the agreed Sendai Framework and collaborate on lowering disaster risk by increasing DRR investment and safeguarding people and property for long-term sustainability. Many disaster preparedness and response (DRR) initiatives including disabled persons, senior citizens, women, and children are now being carried out in Bangladesh. After adopting SFDRR, the government of Bangladesh hosted the first-ever global conference on disability-inclusive disaster risk management in 2015. This event is also known as the Dhaka Conference. The Chair's Summary from GP2017 in Cancun strongly encourages all nations to adopt the Dhaka Declaration, which was adopted at that meeting and subsequently approved by AMCDRR 2016 in New Delhi. In addition, in May of 2018, Dhaka played host to the 2nd International Conference on Disability Inclusive Disaster Risk Management. Even during AMCDRR 2018, held in Mongolia's capital of Ulaanbaatar, the results of the "Dhaka Declaration+" were acknowledged.

2.6 DRR—A Community-Based Approach

The goals of disaster risk reduction (DRR) projects can be to lessen the impact of disasters (through land use zoning or structural measures), make people less at risk of being negatively affected by disasters (through safer housing or more income streams), or both (savings, access to credit, insurance). Community-based risk assessment, awareness building, early warning systems, diversifying, and reinforcing means of subsistence are all examples of non-structural DRR initiatives. Development agencies have a distinct advantage in this respect because of their strategy of collaborating with CBOs and the priority they place on strengthening community capacities.

It is important to note that the term "community-based organization" encompasses a broad range of local initiatives with the goal of enhancing people's personal, communal, and societal well-being. Community service, community building, and community mobilization are all terms that are being used interchangeably. Organizing within communities that share a location, a workplace, or a set of experiences or concerns can be seen as an example of community-based organizations functioning as civil society non-profits. However, it is crucial that they intentionally consider a multi-hazard strategy and include risk reduction and resilience-building initiatives in their programming.

2.7 Disaster Risk Reduction and Sustainable Development

Incorporating disaster risk concerns into sustainable development processes is a fundamental approach of the United Nations-sponsored Hyogo Framework for Action 2005-2015, which aims to increase national and community resilience to disasters. One of its five goals is reducing underlying risk factors, which includes environmental, social, and economic activities, although this is the area where the least progress has been made, at least according to government reports. The alternative post-2015 framework intended to successor the Hyogo Framework for Action would benefit greatly from the SDGs' explicit identification of disaster risk reduction in order to spur the substantive work on underlying disaster risk [7].

Hazards, exposure, and vulnerabilities interact to create disaster risk, which may be thought of as the result of a continuous "risk process." Potential losses may increase over time due to the compounding effect of the risk and may not become apparent until a hazard event actually occurs. This is a major departure from the traditional view of disasters as random occurrences of nature [8]. While geophysical hazards themselves are unavoidable, the extent to which they affect individuals, communities, and even entire nations depends on a wide range of micro- to macro-level political, social, economic, and environmental factors. It is imperative that the elements that increase catastrophe risk be understood so that proactive policy and action may be taken to mitigate them.

2.8 Disaster Risk Reduction: Issues and Challenges

Priorities

Since DRR capabilities and resources are inadequate, expecting development in every area would be impractical. Essentially, governments and other organizations must make "investment choices," or pick and choose which parts of DRR to fund and when.

Adding further complexity is the fact that many proposed solutions focus on growth rather than emergency preparedness. In general, the current DRR recommendations avoid dealing with this specific problem.

One approach is to focus in on measures taken only to lessen the likelihood of calamity. In any case, this would make it stand out from other, similar initiatives. Sustainable development is an alternative to risky expansion. By adopting "decisions and actions that are consciously intended and executed to minimize risk and susceptibility, and to improve resistance and resilience to disaster," invulnerable development seeks to lessen people's exposure to potential catastrophes.'

Partnerships and Interorganizational Coordination

Various elements of DRR cannot be handled by a single entity. DRR considers catastrophes to be complex challenges that need a group effort to solve. However, even in the most basic of emergency management scenarios, coordinating the efforts of the many groups that may rush to a disaster zone to provide aid may be challenging. Relationships between different kinds of organizations and across different sectors (public, commercial, and non-profit, as well as communities) grow significantly more comprehensive and complicated throughout the larger range of DRR. Strong vertical and horizontal links are necessary for DRR (central-local relations become important).

DRR and Governance

In order to implement DRR, the government's traditional role in disaster prevention must be rethought. There is widespread consensus that national governments should play a central role in disaster risk reduction (DRR) because they are responsible for the safety of their citizens, have the resources and capacity to implement DRR on a large scale, are tasked with coordinating the efforts of others, and draft the necessary policies and laws. There must be harmony between these many plans and initiatives [9]. More investigation into the factors that make certain governments better equipped to deal with disasters is required. What motivates people to alter their policies and procedures is still a topic of debate.

Accountability and Rights

Honest collaboration and involvement in DRR are founded on the accountability concept. It is applicable to both publicly responsible governmental institutions and privately funded, non-profit and charitable organizations. Taking responsibility for one's actions is a relatively recent concern in the field of disaster assistance. Those who are most susceptible to and impacted by risks should be the primary targets of accountability. A rights-based approach is being adopted by an increasing number of international assistance and development agencies. Human rights (i.e., those commonly recognized via international accords) and additional rights that an agency considers should be accepted as human rights fall under this umbrella.

2.9 Stages of Disaster Risk Management

All the things that can be done before, during, and after a catastrophe to mitigate its effects and get back on your feet are collectively known as "disaster risk management." Management of catastrophe risk consists of three main phases, as follows (Fig. 2.3):

Before a Disaster (Predisaster): The measures conducted before catastrophe strikes are called "predisaster actions," and their purpose is to lessen the impact on lives and property. Things like educating the public and shoring up vulnerable infrastructure and laying up strategies for how to deal with disasters on a personal and communal level are all important. Activities like these, conducted during this phase, are known as "mitigation" and "preparedness" and are aimed at lowering the risk level.

During a Disaster (Disaster Occurrence): Efforts to alleviate victims' hardships and provide their material need are among these. Activities performed at this level are referred to as emergency response activities.

After a Disaster (Post-disaster): There are measures performed right after a tragedy occurs that aim to hasten the communities' road to recovery and reconstruction. We refer to these processes as "response" and "recovery." The DRMC graphic

Fig. 2.3 Cycle of disaster management with nature of responses. *Source* Coppola [8]

emphasizes the many efforts made throughout the emergency response and recovery phases of disaster management. Tasks like coordinating and providing continuous support are common to both phases, while some are exclusive to one or the other (e.g., early warning and evacuation during emergency response; and reconstruction and economic and social recovery as part of recovery). The DRMC also stresses the importance of the media, noting the correlation between coverage and financial support. In contrast to rapid-onset catastrophes like floods, earthquakes, bushfires, tsunamis, cyclones, etc., the transition into the emergency response stage of a slow-onset disaster like drought is less easily identifiable.

2.10 Defining DRR and Relevant Issues

Disaster Risk Reduction

DRR is often interchanged with hazard mitigation. Predisaster actions are those carried out before a catastrophe strikes in attempt to mitigate its effects. This is a narrower word than "disaster management," which encompasses a wider range of activities, including prevention and response. "The conceptual framework of elements considered with the possibilities to minimize vulnerabilities and disaster risks throughout a society, to avoid (prevention) or to limit (mitigation and preparedness) the adverse impacts of hazards, within the broad context of sustainable development," as defined by the UN/ISDR. For DRR to be properly defined as a dependent variable, the following topics arise for debate.

Risk

In the context of catastrophes, risk refers to the likelihood of adverse outcomes stemming from a hazard occurrence, such as loss of life, property damage, injury, or loss of means of subsistence. The risk of anything happening depends on how susceptible the people in the area are to that something happening.

Hazard

Catastrophe occurrence is quantified by hazard, whereas the potential harm caused by a disaster is quantified by vulnerability. In this way, the likelihood of a catastrophe occurring or the damage it may do could be lessened. Diverse DRR strategies exist, each addressing a particular facet of danger. The incorporation of indigenous knowledge into DRR policy, especially in educational settings, is another strategy for decreasing disaster risk and protecting communities from natural disasters.

Vulnerability

The word "vulnerability" refers to a population's propensity to be negatively affected by risks. The "vulnerability approach," which emerged in the 1970s and 1980s, introduced the idea into the catastrophe discourse by rejecting the notion that natural, environmental factors are the only cause of disasters. It maintains that catastrophes are "normal" and that the vulnerability of a civilization is the primary cause of disasters. Due to the fact that vulnerability is established by social structures and the distribution of power at the regional, national, and international levels rather than by the state of the natural world, it is hard to eradicate vulnerability from the social fabric. There are a wide variety of security holes that might be exploited. The United Nations International Strategy for Disaster Reduction (UNISDR) defines vulnerability as "exposure to risk," placing equal emphasis on physical, economic, environmental, and social dimensions of vulnerability.

Resilience

Definition: "The innate potential of a system, group, or culture predisposed to a shock or stress to adapt and survive by modifying its non-essential qualities and reconstructing itself" (source). This concept is relatively new to the area of disaster management. It was first used to characterize communities and their link to potential calamities relatively recently, although its origins may be traced back to the 1940s in the fields of psychology and psychiatry. Resilience, in contrast to vulnerability and risk, places an emphasis on strengthening rather than weakening; it goes beyond the concept of lowering vulnerability, which implies "dealing with calamities," and instead shifts focus to adaptation and the development of something more robust [10]. Risk and susceptibility may be mitigated through training for future scenarios. The United Nations International Strategy for Disaster Reduction (UNISDR) defines resilience as "the degree to which an entity can respond to and gain control over future risks by drawing from and adapting to its environment."

Capacity

In disaster risk reduction (DRR), capacity is the sum of all the ways in which vulnerability may be mitigated or impacts mitigated. The strength of a group depends on a number of things. They consist of things like land, government, society, and money. Having the capacity to learn, to lead, or to manage initiatives to reduce catastrophes is also an important aspect of community capacity. Ability is frequently used interchangeably with capacity. A community's capacity is measured by how well it can prevent or respond to catastrophic events. Again, taking indigenous knowledge into account enables DRR plans and procedures to enhance the community's preexisting capability to deal with future calamities.

Sustainable Development

Because of its focus on the future, the idea of sustainable development is crucial to the field of catastrophe management. In order to achieve their respective aims, disaster management and sustainable development must work together. The United Nations has been in the forefront of demonstrating the importance of catastrophe management to sustainable development since the turn of the century. The United Nations General Assembly declared in a resolution in 1989 that natural catastrophes impede the sustainable development of poor nations.

Policy Framework

In order to make it simpler for workers to access and comprehend the information included in different policy documents, a policy framework is often used. The organization's policies may be planned and developed with the use of policy frameworks. It is the government's responsibility as a steward on behalf of the public to protect and preserve any information generated or gathered by any individual employed or contracted by the crown. An organization's policy framework is the overarching collection of guiding principles and long-term objectives that are used to shape operational norms and influence strategic decision-making [11].

The term "policy" or "policy research" may also be used to describe the steps used by an organization to arrive at pivotal choices, such as weighing the pros and cons of several program or budgetary options and ultimately selecting one. Definition: Policies are predetermined plans for achieving certain outcomes at the political, managerial, financial, and administrative levels. To those in the business of public corporate finance, a "critical accounting policy" is one that has a high degree of subjectivity but nonetheless has an important bearing on the company's or industry's financial statements. Both intentional and unforeseen consequences may affect policy.

Intended Effects

The structure and setting in which a policy is implemented significantly affect its intended outcomes. In most cases, organizations adopt new policies in an effort to mitigate a known problem or to increase the likelihood of a desired outcome. One way that businesses attempt to prevent unfavorable outcomes is via their procurement procedures. Some companies' standards stipulate that anything costing more than a specific threshold must go through a formal buying procedure. By making this kind of buying mandatory by policy, the business may save costs and improve efficiency.

Unintended Effects

Side effects and unexpected repercussions are common results of policymaking. A change in policy might have unexpected outcomes since the surroundings it seeks to affect or manage are often complex adaptive systems (such governments, society, and major corporations). A government may decide to increase taxes as a matter of policy in order to boost the total amount of money it collects in taxes. Depending on how much of a hike is implemented, it might discourage people from working and so reduce tax collection, or it could cause capital flight. To reduce the likelihood that a policy would have unintended repercussions, it is common practice for those involved in formulating policy to seek to analyze as many potential areas of policy influence as feasible. Some complex adaptive systems, such as societies and governments, make it difficult to predict all of the outcomes of a policy's implementation.

The policy cycle is a framework used by political scientists to examine the lifecycle of a policy proposal. The term "stagiest strategy" also describes this method. These steps are included in one typical version:

1. Agenda setting (Problem identification).
2. Policy formulation.

3. Adoption.
4. Implementation.
5. Evaluation.

Commonly, a classical technique is used to describe policy cycles. In turn, some postmodern scholars argue that cyclical models are insensitive and unrealistic, instead favoring systemic and more complicated models. They take into account a wider variety of stakeholders in the policy sphere, such as non-governmental organizations, the media, academics, think tanks, businesses, lobbyists, and so on.

Structural and Non-structural Mitigation

With the goal of better coordinating the whole disaster management system in mind, the government of Bangladesh has prioritized both structural and non-structural mitigation measures in light of the new concept of disaster management. The government of Bangladesh has the firm belief that structural mitigation measures must be added to non-structural mitigation strategies in order to alter or lessen certain catastrophe consequences [12]. Bangladesh's disaster management programs provide equal attention to both permanent structures and more ephemeral methods of preparedness:

Institutional Arrangement

The government of Bangladesh has taken several noteworthy initiatives over the past few years to establish national and union-level institutional arrangements for effective and systematic disaster management, which should help to alleviate the suffering of those who have been affected by disasters in Bangladesh. The government of Bangladesh has established a series of mechanisms for council and committees from the national down to the grassroots levels in order to maintain proper coordination among the relevant ministries, departments, line agencies, Local Government Body (LGD), and community people, and to ensure their proper functioning in order to alleviate the suffering of the people. The Standing orders on Disaster (SOD) provide as directives for the systems to function optimally [13].

Powerful bodies, the National Disaster Management Council (NDMC) and the In-Ministerial Disaster Management Coordination Committee (IMDMCC), were established to promote and coordinate risk reduction, preparedness activities, and mitigation measures. They hold meetings twice a year and four times a year, respectively. While the National Disaster Management Council (NDMC) is responsible for formulating and reviewing disaster management policies and issuing directives to all concerned, the Interministerial Disaster Management Coordination Council (IMDMCC) is responsible for implementing the directives, maintaining interministerial coordination, and overseeing the services of the Armed Forces and non-governmental organizations (NGOs) involved in disaster management in the country. The MDMR's Emergency Operations Center (EOC) is housed inside a well-established agency called the Directorate of Relief and Rehabilitation (DRR) [14].

Under the direction of the Ministry of Disaster Management and Relief (MDMR) / IMDMCC, the DRR responds during post-disaster emergency situations and runs relief operations for distribution to outlying field levels. To carry out its specialist functions and ensure coordination with line departments/agencies and NGOs, the MDMR has a small, dynamic professional unit known as the Disaster Management Bureau (DMB), which organizes and convenes meetings of the Disaster Management Training and Public Awareness Building Task Force (DMTATF), the Focal Point Operational Coordination Group on Disaster Management (FPOCG), the NGO Coordination Committee on Disaster Management (NGOCC), and the Committee for Rapid Dissemination of Disaster Information.

Community-Based Disaster Management Practices in Bangladesh

The increasing abnormality of various natural disasters raises fundamental questions about the role of the government and how it would manage the disaster, as well as what would be the more rational and effective way of management, in light of the emergence of a changing role for the government and the increasing participation of the public in the government's sociopolitical activities. Policymakers, experts, and professionals are increasingly in agreement that the government cannot and will not adequately manage and handle all types of disasters with its machinery, and that the people in any region of a country must take an active role in responding to and recovering from such events.

Important components of disaster management, such as mitigation/prevention, readiness, response, and recovery, are covered by the current framework for disaster management in the nation during normal times. In today's modern world, disaster management is more of a spectacle than a growing field. Local capacity and preparation measures may be strengthened by community engagement in program design and implementation, as well as through linkage with current development projects.

Adaptative Governance in Disaster Management

Key characteristics of a governance system that promotes growth and aids in risk reduction are defined as accountability, involvement, predictability, and transparency. Reducing the likelihood of catastrophic events is an aspect of governance. Making disaster risk reduction (DRR) a guiding concept in all important development sectors depends on three things: public awareness, political will, and adequate ability. UNDP is pushing forward the theoretical and applied work on DRR governance and mainstreaming thanks to its extensive background in DRR, democratic governance, and development.

Society's coping abilities can only be safeguarded by means of adaptive government. Governance determines the manner in which national and subnational players (including governments, legislators, public officials, the media, the corporate sector, and civil society groups) are willing and able to coordinate their efforts to manage and decrease disaster-related risk. It takes political will to develop policies and provide adequate resources, and enough public knowledge to detect and confront risk. In order to manage and coordinate the actions of key sectors and to account for vulnerable and impoverished populations, processes institutions must have appropriate management and coordination competence.

In reality, there is a cyclical relationship between underdevelopment and disaster vulnerability, since catastrophes not only result in significant damage to capital assets but also impede production and the flow of products and services in the impacted economy, leading to a drop in income. Such effects may have significant long- and short-term consequences for a country's economic growth, including negative effects on GDP, public finances, international trade, and price indices, all of which can exacerbate already-worsening poverty and debt. Many of the poor in the world's least and least-developed nations (LLDCs) rely on the agricultural sector for their livelihood and survival; therefore, any disruption to that industry may have devastating impacts on their lives.

Rising poverty rates, in turn, enhance vulnerability to disasters: when the poor seek to exploit natural resources for subsistence, damaging forests, soils, wetlands, and water supplies, the danger of catastrophes such as landslides rises inescapably. Repeated exposure to catastrophes may lead to chronic poverty because repeated economic strains induced by natural disasters generate income volatility, prompting families to forgo prospective investment. People's exposure to risk may rise if economic factors compel them to settle in high-risk places like flood plains or precarious hillside locations.

Climate Diplomacy

There is rising concern throughout the globe that placing climate change in a larger foreign policy framework is necessary to achieve the agreement and commitment necessary to take action. Stability and security are seen as conventional foreign policy's end objective since they provide the groundwork for human flourishing, international peace and prosperity. Traditional diplomatic tools are becoming less effective in today's interconnected globe.

When greenhouse gas emission is not the goal of anyone "unreceptive" influence, it might be difficult for established alliances and processes to be successful against a problem like climate change. Foreign policy needs fresh thinking to handle the climate change problem, thinking that places climate change involvement outside the milieu container. Even while climate scientists and diplomats or officials from the foreign ministry don't always share a same language, climate diplomacy forces them together.

The global and domestic policy and political connections inherent in environmental diplomacy are also prominent features of climate diplomacy. Successful environmental diplomacy requires a cooperative, multi-lateral approach, which may be revitalized via healthy rivalry for the mantle of international environmental leadership. Trade and investment, development and human rights, and even military security are all traditional sectors of international relations that are increasingly interwoven with environmental concerns.

To help supplement national efforts in preventing and adapting to climate change, climate diplomacy aims to persuade other governments and/or global organizations to adopt certain policies. The issue of climate change is quickly becoming a focal point of international diplomacy at the highest levels.

Indigenous Knowledge in Disaster Management

Although defining indigenous knowledge may be challenging, it often refers to the accumulated and evolving body of expertise within a certain community. It has been used for centuries and modified to fit the local culture and environment and is thus grounded in the experiences of the people who live there. Integration of indigenous knowledge based on universally applicable concepts reflected in locally appropriate behaviors is essential for disaster risk reduction strategies. The requirements for incorporating this kind of knowledge into disaster policy have been established via the idea of Transferable Indigenous Knowledge, which was developed by the Disaster Reduction Hyper base Initiative. In addition, a goal of the post-2015 framework for disaster risk reduction is to use traditional and local knowledge with relevant and suitable scientific information in disaster risk assessment and the creation and implementation of policies, plans, and programmes. Adapting to specific vulnerabilities and the specifics of a certain country's geographical location may greatly improve and fortify the country's overall capacity for adaptation. It's a concept that's gaining traction in the area, as people are beginning to understand that it may help them make better use of resources like sand, gram, mangroves, coral reefs, and stones that are already in their immediate vicinity.

References

1. Cardona OD (2004) The need for rethinking the concepts of vulnerability and risk from a holistic perspective: a necessary review and criticism for effective risk management. Earthscan Publishers, London
2. UNISDR (2006) HFA for DRR. UNISDR, Geneva
3. UNISDR (2007) DRR reporting for action. UNISDR, Geneva
4. Hewitt K (1999) Regions of risk: a geographical introduction to disasters. Harlow, Essex, Longman
5. SDMC (2009) HFA progress report for disaster risk reduction. SDMC, New Delhi
6. Peters K (2018) Accelerating sendai framework implementation in Asia. ODI, London
7. Medina JC (1992) Community awareness and participation programme in disaster preparedness and rural development. UNCRD Press, Nagoya
8. Coppola D (2007) Introduction to international disaster management. Elsevier, Burlington MA
9. ADB (2004) Disaster and emergency assistance policy. Asian Development Bank, Manila
10. ADPC (2008) Mainstreaming disaster risk reduction into development policy, planning and implementation in Asia. Bangkok, ADPC
11. Benson C, Twigg J (2004) Measuring mitigation: methodologies for assessing natural hazard risks and the net benefits of mitigation. Switzerland, Geneva
12. Holloway A, Pelling M (2006) Legislation for mainstreaming disaster risk reduction. Teddington Publishers, UK
13. DMB (2008) Disaster management mechanism in Bangladesh. Government of Bangladesh, Dhaka
14. MoDMR (2009) Disaster emergency coordination system. Dhaka, MoDMR

Chapter 3
Disaster Management in Bangladesh: Historical Perspective

3.1 Introduction

Bangladesh (also known as East Pakistan and East Bengal) has a long history of natural disasters that has been widely publicized over the last two centuries. From the commencement of British administration till the present day, we have seen signs of big natural calamity occurrences occurring on this country. While emergency medical care was always a priority, the British never developed a comprehensive strategy for disaster management. Upon further investigation, we discovered that the literatures we consulted did not address the topic of humanitarian response after a catastrophe. Nation-state consolidation was a challenge for both India and Pakistan after the 1947 Partition of India, but Pakistan's East and West were "separated by almost 1000 miles of Indian territory," compounding an already challenging situation [1].

Since gaining independence, she has built up a formidable domestic capability dedicated to disaster prevention and relief efforts. Bangladesh is more at risk from a variety of natural catastrophes due to its location and topography, both of which render the nation low-lying and easily submerged. The reasons why Bangladesh is at risk of natural disasters all year long are laid forth in Figs. 3.1 and 3.2.

Since its inception in 1955 as a result of Pakistan's One Unit plan, the province of Pakistan known today as Bangladesh has occupied the Bengal area in the northeast of South Asia. Area-wise, East Pakistan was 147,570 km (56,977 miles) in size. It was bounded to the east, north, and west by India, and to the south by the Bay of Bengal. As one of Pakistan's provinces, East Pakistan had a sizable portion of the country's population, a sizable political presence in the federal government, and a sizable economy [2] (Fig. 3.3).

This chapter focuses on the historical development of disaster management in Bangladesh from British period to Pakistan period.

M. T. Islam, *Disaster, Governance and Development*, Disaster Studies and Management,
https://doi.org/10.1007/978-981-99-1631-3_3

- One third of population below the poverty line and 17% or some 27 million people still live in extreme poverty
- Sea level rise has the potentials to displace nearly 30 million people living in the coast
- in terms of people exposed to Bangladesh is ranked globally:
    - 1st for floods, 3rd for tsunamis and 6th for cyclones
        - 14% GDP exposed to disasters per year – the highest ranking in the world
        - Between 1980-2008:

Fig. 3.1 Disaster and climate profile of Bangladesh. *Source* United Nations Development Programme, climate change, environment and disaster profile of Bangladesh, 2012

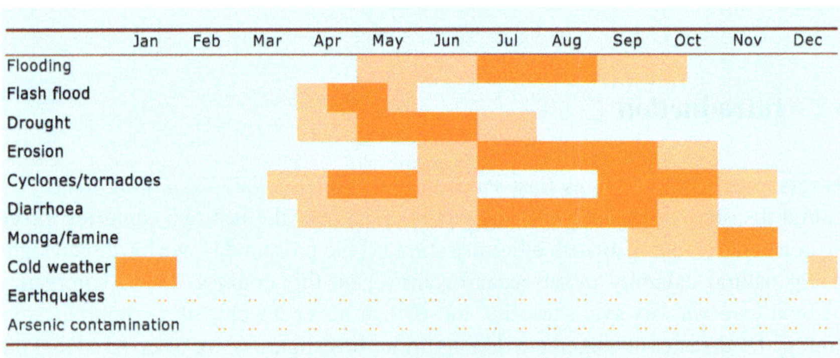

Fig. 3.2 Disasters calendar of Bangladesh. *Source* World Food Programme Bangladesh, 2011

Fig. 3.3 Map of Pakistan with East Pakistan highlighted in red color

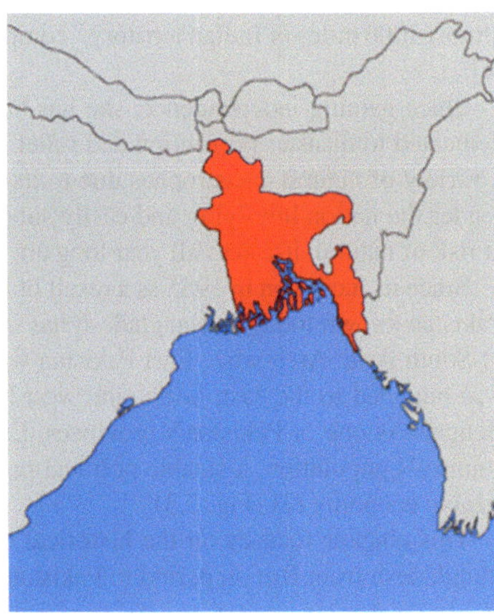

3.2 Disaster Management During British Rule

The era of British authority in the Indian subcontinent often referred to as "the British Raj" spans from around 1858 till 1947. Areas directly managed by the UK and the princely kingdoms controlled by individual rulers under the auspices of the British Crown were both part of the territory popularly referred to as India today. A corporation sole, The Crown is the personification of executive, legislative, and judicial power in the Commonwealth dominions and their constituent province and state governments.

It originated in Britain as a way to decouple the monarch's personal wealth and power from the state's crown and property. British colonialism helped disseminate the idea, and it is now well established in the legal language of the other 15 sovereign countries. Occasionally, the territory was also referred to as the Indian Empire. It formed both the League of Nations and the United Nations and sent athletes to the Olympics as India in 1900, 1920, 1928, 1932, and 1936 [3].

This form of government was established in 1858, when Queen Victoria (who in 1876 was proclaimed Empress of India) took over from the British East India Company, and continued until 1947, when the British Indian Empire was divided into the Union of India (later the Republic of India) and the Dominion of Pakistan (later the Islamic Republic of Pakistan, and the Eastern half of which, still later, became the Republic of Bangladesh). Upper Burma was added in 1886, and the resultant union, Burma, was managed as a province until 1937, at which point it became a distinct British colony that earned its own independence in 1948 [4]. Lower Burma was already a part of British India at the start of the Raj in 1858.

As the twentieth century began, British India was split into eight regions, each of which was governed by a governor or lieutenant-governor. Among them are Bangladesh and Bengal. The Indian Army, together with the smaller but highly skilled Indian Civil Service, was funded out of the Raj's overall budget. As a result, the British government never established a specific agency to handle emergencies like famine or natural catastrophes or allocated sufficient funds to do so.

The Great Famine of 1876–1878 and the Indian famine of 1899–1900 were two of the greatest famines ever documented; 6.1–10.3 million people perished during these two events alone. British officials started discussing famine strategy immediately after the disastrous relief attempt during the Orissa famine of 1866, and in early 1868, Sir William Muir, Lieutenant-Governor of the North-Western Provinces, issued a famous order declaring that.

As stated in the document, "every District officer would be held personally responsible that no deaths from starvation occurred which could have been avoided by any exertion or arrangement on his part or that of his subordinates," implying that the governor of the province will be responsible to distribute the necessary relief assistance among the disaster victims in order to help them rebuild their lives and livelihoods [5].

During the British era in the Indian subcontinent, the food department was tasked with coordinating relief efforts with the local government unit. The famine in Bengal during 1943 was poorly handled by the British administration. The British administration in Bengal failed miserably in 1943 to control the famine because of a combination of factors, including a lack of knowledge about the severity of the situation, insufficient preparation, a shortage of relief supplies, inadequate management, and unequal distribution of aid.

3.3 Causes and Impact of the Bengal Famine 1943

After the Japanese occupied Burma in WWII, the Bengal Province of British India (modern-day West Bengal, Odisha, Bihar, and Bangladesh) experienced a famine in 1943. Starvation claimed the lives of almost 3 million people. Death toll estimates from hunger, malnutrition, and illness in Bengal's 60.3 million strong population range between 1.5 and 4 million. When food finally arrived in December 1943, half of the victims perished from starvation and the other half from sickness [6]. As a result of the British Empire seizing 60% of all crops and forcing Bengal to furnish a bigger share of the food for their army to fight the Japanese, the demand for food much outpaced the supply even though food output was higher in 1943 compared to 1941. The biggest death toll was not among the really poor, as was the case in past Bengal famines, but rather among craftsmen and small merchants whose revenue evaporated when people spent all they had on food and no longer hired cobblers, carpenters, etc. Millions of households were wiped out economically and socially as a result of the famine.

India's food supply was unstable at the outset of World War II due to a number of crop failures and isolated famines, but the situation was brought under control by the implementation of the Indian Famine Codes. A small-scale famine occurred in Bengal in 1940–1941, although widespread mortality was avoided thanks to swift government response. The national spike in food costs prompted the Indian government to call for meetings with state and provincial authorities and to establish price controls. The famine was brought on by a combination of dwindling supplies and rising demand.

The winter 1942 Aman rice harvest was initially predicted to be poor or mediocre, then in October, a storm and three tidal surges devastated the region, further reducing the yield. Wind and heavy rain ravaged an area of 3200 square miles, while tidal surges rushed over 450 square miles. Farmers, shoppers, and merchants all lost their reserves. The disaster resulted in the deaths of 14,500 humans and 190,000 animals. About 2.5 million Bengalis had their homes, livelihoods, and property destroyed or damaged [7].

With food supplies dwindling, the famine was exacerbated by the British empire's decision to demolish arable land in Bengal to make space for opium poppy production for export. Right-wing British measures that exacerbated the famine included mandating that farmers grow indigo instead of rice and outlawing the "hoarding" of food. This stopped merchants from stockpiling food and other supplies to help people through lean times. Since 1941 was such a short year, it was said that Bengal did not have the customary amount of stored food from the previous harvest to utilize as emergency food (as they certainly did when the December 1943 crop was harvested) [8]. Since the customary excess supplies did not get carried over from the strong December 1941 harvest into 1943, this meant that the war effort had to rely on rationing.

For the last ten years, Bengal has had to rely on food imports. Calcutta was generally supplied from Burma. After the British Empire was soundly defeated by the Japanese military in 1942 at Singapore, the Japanese military immediately invaded Burma. During the years between the wars, Burma was the world's leading rice exporter. Even though Burma only supplied 15% of India's rice in 1940, a somewhat larger percentage of rice in Bengal came from Burma because of the province's closeness to the country. Bengal and the other portions of India and Ceylon, who relied on Burma [8] as a food source, had to go elsewhere after the Japanese conquest of Burma in March 1942.

However, starvation struck the Eastern cities of Madras, Orissa, and Bengal, as well as the Western cities of Cochin, Trivandrum, and Bombay. The burden of supplying the remainder of India and Ceylon rested on the few excess provinces, namely the Punjab. A similar increase in demand for food in Bengal may be attributed to the migration of Burmese refugees. Increased local demand in Bengal (up to 200,000 tons of grain imported, plus an unknown amount of grain, and a lot of fresh food purchased in Bengal) likely did not boost overall food consumption in India, but it did have a substantial impact on the region [9].

Lack of Statistics

The inability to detect and respond to the famine was largely attributable to a lack of data. Administrators and statisticians in India were aware of how inaccurate their estimates of agricultural productivity were well before the famine. "Totally untrustworthy" and "useless for any purpose" were two phrases used to characterize the figures. It was also claimed that "no dependable statistics existed in Bengal," that "what statistics existed were disbelieved by the very Government which issued them," and that there were "no meaningful production statistics," with the averaging procedure only amplifying the error. Finally, senior officers changed the calculated figures at will, with roughly half of the estimates undergoing some sort of revision [10].

The land tax system in Bengal made it difficult to collect accurate agricultural data. In 1942, a revenue officer would estimate the area planted and the probable yield for a 750,000 acre (310,000 ha) area in order to provide a crop forecast. These estimates were then aggregated and "adjusted" by successive levels of Department of Agriculture officials.

Food officials used a variety of other estimates, cross-checking them against observable facts. This included information revealed in mail censorship, such as letters from farmers, landlords, and traders on crop yields, reports from Special Branch (the secret police), reports from other departments, etc. Traders acted on their belief in a serious shortage and made a lot of money.

Analysis based on the level of the production forecast and, in particular, on year-to-year differences in production forecasts is precluded because the number of people who needed to be fed was unknown. The Census of Population was known to be unreliable at best, and the 1941 Census was particularly unreliable.

Provincial Government Inaction

No of what caused the famine, it was only the arrival of food supplies from other parts of India that kept people from starving to death. In previous regional famines, the Indian government had given money to the hungry and allowed the trade to bring in grain, which helped those areas. However, this had disastrous results in Orissa in 1888, when the shortage was not regional but national, and so the government did not repeat this strategy in 1943. Provincial administrations that had been duly elected in the previous year imposed trade restrictions with federal approval in 1942. Politicians and bureaucrats in excess regions like the Punjab enacted restrictions to stop food from leaving their provinces for the famine zones of Bengal, Madras, and Cochin.

Because of the potential for civil upheaval, it was important to ensure the food security of both the local population and the populations of nearby provinces. Politicians and bureaucrats obtained power and patronage, and the ability to extort bribes for shipping licenses. The price of advertising and doing business shot far up. The market could not get grain to Bengal, however lucrative it may be. The major commerce route, established for hundreds of years, was up the river system and thus stopped to function, leaving the railway as the sole option of transporting food into Bengal. Calcutta, the second largest metropolis in the world, barely had two weeks' worth of food on hand in March 1943 when grain deliveries ceased [11].

After realizing its error, India's government reinstated free commerce. Provincial populations shrank. The Government of India Act of 1935 had stripped the Government of India of most of its authority over the provinces, forcing them to rely on negotiation, but "once again, the Government of India misjudged both its own influence and the temper of its constituents," who had gone too far by this time to pay much attention to the centre.

Thus, politicians, civil officials, local government authorities, and police impeded the flow of food to famine regions despite the Government of India decreeing that there should be unrestricted commerce in grain. In other situations, provinces confiscated food in transit from other provinces to Bengal. According to Mahesh Chandra: "But persons like Bhai Permanand indicate that despite many businessmen wish to transport grain [to Bengal], the Punjab Government would not issue them licences.... He testified to vast amounts of undisposed-of rice existing in the Punjab." This was said in 1943.

Eventually, when the new Viceroy, the successful military Archibald Wavell, was ready to enter office, there was a clear threat by the Government of India to compel the elected governments to deliver grain. We saw the first large-scale shipments of grain to Bengal. The government of Bengal was sluggish to begin aid efforts, and at one point in 1943, it restricted aid to preserve money even though it was easily available. Those who support any of the two competing Bengali governments, A. Both the administration of K. Fazlul Huq (in office from December 1941 to March 1943) and the government of Khawaja Nazimuddin's Muslim League (in office from

April 1943 to March 1945) blamed the other for the disaster because of the inactivity and corruption of the other. Premier of Bengal and a member of the A. K. Fazlul Huq, who had foreseen the impending famine, was removed when his warnings fell on deaf ears [12]. Critics pointed to "the feebleness of its moral and administrative norms" in the administration, which had done practically little to prepare for the famine.

Motives Behind the Failures in Relief Action

There is evidence that the elected Provincial governments, their public employees, and some prominent members in the Indian civil service thought, or at least claimed to think, that Bengal had sufficient food supplies that could be made accessible with efficient management. No credible production data existed to back this position, and the data that did exist was "hopelessly faulty" and pointed to a severe deficit. People in Bengal are hungry, yet others say the crisis was caused by hoarding and should be addressed administratively rather than via food relief.

Even at the height of the famine in November, the Director General of Food in the Council of State said that "the major trouble in Bengal has been not so much an intrinsic shortage of essential food grains as a breakdown of public confidence." This is despite the fact that the Government of India admitted there was no intrinsic shortage in Bengal in the spring of 1943 [13].

At the height of the famine, on October 19, 1943, Wavell wrote in his notebook, "On the food situation Linlithgow [The outgoing Viceroy] states primary factor morale [i.e. panic hoarding]." If there had been sufficient food and water, the famine would not have been so severe [14]. The wealthiest 10% of Bengal's population, the only ones who could afford it, would have had to stockpile enough rice to last two years, on top of the two years' worth they had already stored up, and keep it hidden away until the conclusion of the war, while their neighbors went hungry. This is substantial evidence against the stockpiling theory, since there was never any hint that anything of the type occurred.

Administrative and Policy Failures

Many administrative, civil policy, and military policy errors are documented in The Famine Inquiry Commission (1945). No other nation hit by hunger has released similar complaints of its own government's response. The primary mistake was not preparing for rationing in 1939 by establishing a food administration. There was also the inability to implement a uniform food policy throughout all of India that required everyone to make the same sacrifices. The administrative safeguards should preclude any kind of interference in the absence of this.

The failure of political and administrative authorities to implement a plan to confiscate food stockpiles in overabundant regions also contributed (it was acceptable in deficit areas). However, many other regulations (such as the Boat Denial Policy, the Rice Denial Policy, and numerous buying strategies) contributed to local shortages or raised the mortality rate, but they did not cause the famine itself. One official who participated in the Famine Inquiry Commission's probe said, "We had trouble about one issue." It was because there wasn't a centralized government in charge of dealing with the hunger.

In December of 1942, there was a severe rice scarcity in the city of Calcutta. Government efforts therefore centered on acquiring regional excess stocks with the goal of supplying Calcutta. To combat the famine, the government allowed unrestricted commerce in rice in Eastern India in the hopes that merchants would sell their supplies to Bengal. However, this strategy also failed to successfully transfer huge inventories to Bengal. The high food costs in April and May were the result of a marketing campaign designed to persuade the public that the shortage of food did not justify the prices.

In the wake of the failure of these propaganda campaigns, there was a mad scramble to find hidden supplies. H. S. Suhrawardy, Bengal's Minister of Civil Supplies from April 24, 1943, stated that there was no scarcity of rice in Bengal and launched a campaign of scaring "hoarders": this generated theft, extortion, and corruption but did not increase the quantity of grain on the market. It became clear to the administration that the magnitude of the supply loss was greater than they had previously thought when these efforts repeatedly failed to discover big supplies.

3.4 Disaster Management During East Pakistan

From 1947 to 1971, a Pakistani province called East Pakistan existed in the Bengal area of northeastern South Asia as a result of the One Unit project. During the British Raj, Bengal was split in two, with the Eastern half being mostly Muslim and the Western half being predominantly Hindu. Separation between Bengali Muslims and Hindus was exacerbated by the partition of Bengal, which witnessed the widespread recurrence of Hindu–Muslim riots. Following the 1947 approval of the 3 June Plan given by Viceroy of India Lord Earl Mountbatten, Muslim majority areas of Bengal backed the partition and united into the new province of East Bengal of the Dominion of Pakistan [15]. The following catastrophes happened in East Pakistan (Table 3.1):

When disaster strikes, the Food Department and the Health Services Unit will step in to provide emergency aid. Even more so, it was found that local government entities

Table 3.1 Major disasters occurred during Pakistan period

Disaster event	Year of occurrence	Number of death
Cyclone	1960	15,000
Cyclone	1961	12,000
Cyclone	1963	12,000
Cyclone	1965	20,000
Cyclone	1966	15,000
Cyclone	1970	206,000

Source Data collected and compiled by the Researcher

had a symbolic function in aiding those affected by natural disasters. Response and recovery services were negligible, and the quality of such services was terrible. Pakistan's weak disaster management system may be traced back to the country's weak leadership, outdated policies, and ineffective laws. Moreover, the Pakistani military was sent to aid with the search-and-rescue efforts.

As an example of catastrophe management gone wrong, we may look at what happened in Pakistan in the 1970s. On November 12, 1970, a destructive tropical storm known as Bhola hit what was then East Pakistan (now Bangladesh) and the Indian state of West Bengal. It was one of the worst natural catastrophes in recent history, and the deadliest tropical storm ever recorded [16]. The storm surge that swamped many of the low-lying islands of the Ganges Delta was responsible for up to 500,000 deaths. To top it all off, this cyclone was the strongest of the whole 1970 North Indian Ocean cyclone season.

Intensifying as it moved, the cyclone that began developing over the middle of the Bay of Bengal on November 8 eventually reached its northernmost location. On November 11, its peak winds reached 185 kph (115 mph), and the next day it made landfall in East Pakistan (now Bangladesh). Many of the outlying islands were completely wiped away by the storm surge, and crops were ruined all throughout the area. Over 45 percent of Tazumuddin's 167,000 residents perished in the disaster. As the storm's aftermath dragged on, the Pakistani government headed by junta leader General Yahya Khan came under fire from both domestic and foreign political figures in East Pakistan for its handling of relief efforts.

The opposition Awami League won the province by an overwhelming margin in the election held a month later, and the subsequent unrest between East Pakistan and the central government sparked the Bangladesh Liberation War, which resulted in widespread atrocities and ultimately led to the formation of the independent nation of Bangladesh. The storm and the subsequent conflict prompted former Beatle George Harrison and Bengali musician Ravi Shankar to create the first-ever charity concert for Bangladesh in 1971.

The 1970 Bhola cyclone was a destructive tropical storm that affected then East Pakistan (present-day Bangladesh) and India's West Bengal on November 12, 1970. It was one of the worst natural catastrophes in recent history, and the deadliest tropical storm ever recorded. The storm surge that swamped most of the low-lying islands of the Ganges Delta was largely responsible for the estimated half-a-million deaths that the storm caused.

This cyclone was the sixth cyclonic storm of the 1970 North Indian Ocean cyclone season and also the seasons strongest. Intensifying as it moved, the cyclone that began developing over the middle of the Bay of Bengal on November 8 eventually reached its northernmost location. On the afternoon of November 12, it made landfall on the coast of East Pakistan (now Bangladesh), having achieved its peak the previous day with winds of 185 km/h (115 mph). Many of the islands in the vicinity were completely wiped off by the storm surge, together with their populations and their harvests. Over 45 percent of Tazumuddin's 167,000 residents perished in the disaster.

Political leaders in East Pakistan and the foreign media criticized the Pakistani government headed by junta leader General Yahya Khan for its sluggish reaction to the storm's aftermath. A month later, the opposition Awami League won an overwhelming victory in the province's elections, and the continuing unrest between East Pakistan and the central government sparked the Bangladesh Liberation War, which led to widespread atrocities and ultimately the formation of the country of Bangladesh. In 1971, in response to the storm and the conflict, former Beatle George Harrison and Bengali musician Ravi Shankar organized The Event for Bangladesh, the first-ever charity concert.

Causes

On November 5, the last of the remnants of Pacific Tropical Storm Nora, which had lingered for two days in the South China Sea, headed west across the Malay Peninsula. On the morning of November 8, the remains of this system helped birth a new depression in the central Bay of Bengal [17] (Table 3.2).

While slowly heading north, the depression strengthened, and the following day the India Meteorological Department classified it as a cyclonic storm. No nation in the area had a tradition of assigning names to tropical storms at the time, therefore none were given a new name. In terms of human lives lost, this cyclone event is among the worst in recorded history. Right-hand graph displays the same data.

Later that night, the storm stalled out about 14.5° N, 87° E, but on November 10, it picked up speed and headed north. On November 11, while the storm was making its way into the head of the bay, it gained strength, becoming a severe cyclonic storm. Later that day, when it had a clear eye, sustained winds of 185 km per hour (115 miles per hour) were recorded during a three-minute period. High tide in East Pakistan occurred at the same time that the hurricane made ashore that evening of November 12. The system started to deteriorate as soon as it made landfall, and by the time it was around 100 km (62 mi) south-southeast of Agartala on November 13,

Table 3.2 Deadliest tropical cyclones

Rank	Name/Year	Region	Fatalities
1	Bhola 1970	East Pakistan (Bangladesh)	500,000
2	*India 1839* (https://en.wikipedia.org/wiki/Pre-1980_North_Indian_Ocean_cyclone_seasons#November_1839_Coringa_cyclone)	India	300,000
	Haiphong 1881 (https://en.wikipedia.org/wiki/1881_Haiphong_Typhoon)	Vietnam	300,000
4	*Nina 1975* (https://en.wikipedia.org/wiki/Typhoon_Nina_%281975%29)	China	229,000
5	*Nargis 2008* (https://en.wikipedia.org/wiki/Cyclone_Nargis)	Myanmar	140,000

Sources NOAA (http://docs.lib.noaa.gov/noaa_documents/NOAA_related_docs/death_toll_natural_disasters.pdf), MDR (http://www.metta-myanmar.org/publications/Nargis%20RRD%20Final%20Report.pdf)

it had been downgraded to a cyclonic storm [18]. The storm quickly lost its strength that afternoon, becoming a residual low in southern Assam by that evening.

Preventive Action by the Government

Many ship reports were sent to the Indian government from the Bay of Bengal with meteorological details on the cyclone, but due to the typically unfriendly ties between India and Pakistan, this data was not shared with the Pakistani government. Many people were apparently caught off guard by the storm. It seems that the storm warning system in East Pakistan was not used adequately, which may have resulted in the loss of thousands of lives. On November 12, the Pakistan Meteorological Department published a bulletin warning of impending danger along the shore. It was said that Pakistan Radio blasted a "huge danger signal" as the storm approached the shore. Those who made it out alive afterward said it didn't signify much to them, but they did realize that the top warning symbol signified danger was imminent.

After two devastating storms hit East Pakistan in October 1960, killing at least 16,000 people, the Pakistani authorities reached out to the United States government for help in creating a mechanism to prevent similar disasters. In 1961, National Hurricane Center director Gordon Dunn conducted an in-depth study and presented his findings. While many of Dunn's suggestions were implemented, others were not, unfortunately.

Chittagong's weather station, located 95 km (59 miles) to the east of where the storm made landfall, experienced gusts of 144 km per hour (89 miles per hour) until its anemometer was blown off at about 2200 UTC. About 45 min later, a ship moored in the same port reported a high gust of 222 km/h (138 mph). The storm made landfall in the Ganges Delta, producing a storm surge as high as 10 m (33 feet). Chittagong's port had a storm tide that was 4 m (13 feet) above normal sea level, with the surge contributing another 1.2 m (3.9 feet) to the total [19].

There were reportedly no living people on any of the 13 islands in the Chittagong region, according to Pakistani radio. From the air, it was clear that the whole southern half of Bhola Island had been devastated, as had the rice harvests on neighboring islands Hatia and the mainland. There were reports of damage to numerous ships in the ports of Chittagong and Mongla, and water levels at the airports in Chittagong and Cox's Bazar were less than 1 m (3.3 ft) for several hours. There were an estimated $86.4 million in damages caused by the cyclone in 1970 (about $450 million in 2006 dollars) and the cyclone had an indirect impact on almost 3.6 million people [20].

According to those who made it out, almost eighty-five percent of the residences in the region were either completely or seriously damaged. The region's marine fisherman, 90% of whom lost their livelihoods due to the loss of 9000 offshore fishing vessels, also took a major hit. Forty percent of the 77,000 ashore fishermen were killed by the typhoon, and many of those that survived were left badly disabled. In a location where seafood makes up over 80% of the population's protein intake, the storm damaged roughly 65% of the area's fishing capability. Damage to agriculture was also substantial, costing $63 million. Crops were destroyed and 280,000 livestock were killed. More than 75% of the population was still receiving food help from rescue workers three months after the storm, and over 150,000 people are still dependent on aid for more than 50% of their food needs [21].

Intensity of Cyclone70 and Impact

The storm clouds over the Bay of Bengal were observed by American weather satellites as they crossed the sky in a circle of death and destruction. Forms of the yearly cyclone were beginning to take form. As she made her way with murderous purpose into East Pakistan, she was given the codename "Opal," and footage from above television cameras was sent down to viewing locations on the ground (now Bangladesh). Even when radio stations broadcasted warnings, few of the farmers in the Ganges delta (a low-lying triangle of islands and rivers) listened. They expected the annual assault from the water, but they had no idea what kind of fear was gathering up in those furious, offshore clouds. In 1970, on Friday, November 13, 27,000 people were killed in a matter of minutes, and just four out of 4500 homes were later determined to be intact. Kamaluddin was one of the privileged nine thousand people to get it through that dreadful night; for five days, they floated aimlessly in the flooded wastes without food or medical attention.

Rough rafts were built for exactly such an occasion, and some islanders were able to escape. Some hung from bamboo poles, while others clung to trees. The staggering numbers of lives lost due to the cyclone's tidal surge were practically unprecedented. Over three hundred and fifty thousand homes were either destroyed or severely damaged in the Ganges Delta, resulting in the deaths of over two hundred thousand people and the displacement of another two million. Next morning, an area about the size of Greater London was nearly completely destroyed.

Box 3.1: Losses and Damage Caused by Devastating Cyclone 1970

No of people died 220,000

No of people became Injured/wounded 2000,000

No of houses damaged 350,000

Source data collected and compiled by the Researcher

However, the immediate crisis necessitated extensive rescue efforts, and nations throughout the globe reacted by sending aid of every kind. Only West Pakistan, which is stuck between India's Eastern boundaries and the sea, has been noticeably hesitant to react to the plight of her state. Indeed, unlike in the past, West Pakistan did not designate East Pakistan as a disaster area. West Germany offered to set up a 150-bed emergency field hospital staffed by physicians and nurses, but the country declined. A lone Pakistani Air Force helicopter and a couple of small aircraft flew out from Dacca Airport to provide aid to the devastated islands while battered East Pakistan estimated the astronomical cost. They made a feeble attempt, but local authorities were unprepared for even the more common tragedies produced by late-winter storms, much alone anything on this vast a scale.

International Response

Despite their normally bad ties, India was one of the first countries to give help to Pakistan, and by the end of November, they had donated $1.3 million (1970 USD; $6.9 million 2007 USD) to the relief operations. Because the Pakistani government wouldn't let the Indians fly supplies into East Pakistan, they had to be carried overland at a much slower pace. The Indian government also claimed that Pakistani authorities had turned down an offer of help from West Bengal, which included military planes, helicopters, and boats [22].

U.S. Ambassador to Pakistan John Bass promised to "help the East Pakistan administration in any practicable manner," and President Richard Nixon authorized a $10 million ($61 million in 2016) fund to give food and other needed aid to storm survivors. There were a total of six helicopters sent to East Pakistan, two from a Nepalese relief mission and four from the United States [23]. Approximately, 200 thousand metric tons of wheat were sent from the United States to the affected area. By the end of November, a total of 38 helicopters, including 10 British and 10 American, were active in the disaster zone. About 50 small boats were given by the Americans, while the British had provided 70.

A week after the storm struck, CARE stopped sending relief to the nation because they were hesitant to allow the Pakistani authorities to distribute the aid alone. In January, however, they had settled on a plan to build 24,000 homes made of cement brick for a total of around $1.2 million ($7 million in 2016). US officials were concerned that the Pakistani government would take too long to decide how to spend the humanitarian money, so the $7.5 million ($46 million in 2016) that had been approved by Congress in February remained in Washington. The majority of the funds were allotted for restoring homes and erecting cyclone shelters. Despite the United States Peace Corps' best efforts, the government of Pakistan refused to accept any of the volunteers it had given.

In order to aid in the rescue operations, a Royal Navy task force, led by HMS Intrepid and HMS Triumph, departed Singapore and headed toward the Bay of Bengal. They brought in a rescue squad, equipment, and eight helicopters and eight landing boats. In order to assess the damage and begin rescue efforts before the ships arrived, fifty troops and two helicopters were airlifted in. On November 24, the British task force docked off the coast of Pakistan, and the 650 soldiers onboard the ships immediately started deploying landing boats to transport supplies to islands in the area. British Disasters Emergency Committee's appeal for disaster aid in East Pakistan raised around £1.5 million (£21 million in 2016).

Help to the tune of $2 million was promised by the Canadian government. There were $1.3 million worth of helicopters and other equipment supplied from France and West Germany. Pope Paul VI asked people to pray for the victims of the catastrophe and stated that he will visit Dhaka while in the Far East. Later, the Vatican sent $100,000 to help with the relief operations [24]. At the beginning of 1971, there were still four Soviet helicopters in the area, bringing much-needed aid to devastated regions. In the wake of the storm, the Soviet aircraft replaced the British and American helicopters, which had come under fire from local Bengalis.

An army medical mission from Singapore landed in Chittagong, East Pakistan, on December 1, 1971. They were sent to Sandwip, where they helped vaccinate thousands of people and provide medical care to roughly 27,000. On December 22, after delivering over $50,000 worth of medical supplies and 15 tons of food to the storm victims, the expedition returned to Singapore. In December, the Japanese government granted $1.65 million in aid assistance. Prior to this donation, the Japanese government was criticized for its meager contributions to the relief effort.

Even though East Pakistan already had enough supplies on hand, China's first consignment of supplies was a planeload of 500,000 doses of cholera vaccine. Cash in the amount of $1.2 million was handed to Pakistan by the Chinese government. Within a few days of the hurricane hitting, Mohammad Reza Pahlavi proclaimed the tragedy to be an Iranian one, prompting the dispatch of two planes with relief supplies. Numerous smaller, poorer Asian countries contributed little amounts.

UNICEF has started a campaign to collect another million dollars, while the United Nations has provided $2.1 million in food and cash. In the months after the disaster, UNICEF assisted in restoring water supply by fixing over 11,000 wells. U Thant, the UN secretary-general, launched two relief efforts in August to help those affected by the typhoon and the civil violence. According to him, just roughly $4 million had been provided for urgent needs, which was well below the goal of $29.2 million. The League of Red Cross Societies had raised $3.5 million by the end of November to help those affected by the tragedy [25].

Rebuilding the storm-ravaged neighborhood is anticipated to cost $185 million, according to the World Bank. The bank prepared a detailed strategy for the Pakistani government's rehabilitation. It called for bringing homes, water systems, and other infrastructure back to their prestorm condition. It was intended to work in tandem with an existing, comprehensive flood control and development initiative. The bank extended financing in the amount of $25 million to aid in the economic revival of East Pakistan and the building of safe havens for its citizens. The IDA has never before extended rebuilding loans before (Table 3.3).

From Singapore, the British dispatched warships, landing boats, and infantry. Throughout the UK, people reached out to give assistance, and organizations including Oxfam, War on Want, the Red Cross, Christian Aid, and the Save the Children Fund all chipped in £4000 to aid immediately. The United States sent helicopters and cargo aircraft to the scene. West Germany sent 2500 tents, while France did the same [26].

Then something else horrible happened. No large-scale air, land, or river conveyance was available; thus, supplies were stacked up at Dacca Airport, unable to reach the hungry, destitute on distant islands. From the time daylight dawned over the massive floods, the survivors were placed in a position of unimaginable tragedy. They worked non-stop all day long digging graves for the deceased. Children, too, pitched in to help with the grisly job. Cattle that drowned decomposed in the sun, and diseases spread from the contaminated water. Immunizing the survivors against cholera and typhoid was an extreme last resort.

Table 3.3 Figures of external humanitarian assistance to the victims of cyclone 1970

Name of the country/agency	Nature of assistance
West Germany	150-bed emergency field hospital, 2,500 tents
Oxfam	¬£4000 for immediate relief
War on Want	¬£4000 for immediate relief
Red Cross	¬£4000 for immediate relief
Christian Aid	¬£4000 for immediate relief
Save the Children Fund	¬£4000 for immediate relief
India	$1.3 million for relief and rehabilitation assistance
USA	$10 million grant to provide food and other essential relief to the survivors
CARE	Construction of 24,000 cement brick houses at a cost of about $1.2 million
British *Disasters Emergency Committee* (https://en.wikipedia.org/wiki/Disasters_Emergency_Committee)	£1.5 million for disaster relief
Canadian (https://en.wikipedia.org/wiki/Canada) government	$2 millions of relief assistance
France (https://en.wikipedia.org/wiki/France)	$1.3 million for relief support
Vatican	$100,000 for relief support
Government of Singapore	$50,000 worth of medical supplies and 15 tons of food for the victims of the storm
Japan Government	$1.65 million of relief funds
Chinese Government	$1.2 million in cash
United Nations (https://en.wikipedia.org/wiki/United_Nations)	$2.1 million in food and cash
League of Red Cross Societies (https://en.wikipedia.org/wiki/League_of_Red_Cross_Societies)	$3.5 million to supply aid to the victims of the disaster
World Bank	$185 million to reconstruct the area devastated by the storm

Source Data collected and compiled by the Researcher

Government Response in the Aftermath of Cyclone 70

In spite of several hiccups and setbacks, I am confident that everything is being handled and will be completed as promised. Three Pakistani gunboats and a hospital ship carrying medical staff and supplies departed Chittagong for the islands of Hatia, Sandwip, and Kutubdia the day after the cyclone slammed the coast. In the two days after the hurricane made landfall, teams from the Pakistani army visited several of the affected districts. After returning from a state visit to China on November 16, Pakistani President Yahya Khan flew over the affected region. The president demanded that "no effort be spared" in helping those who had been affected. On 21, a week after the hurricane made landfall, he declared a day of national mourning and ordered all flags to be flown at half-mast in tribute to the victims.

In the aftermath of the storm, the Pakistani government dedicated one military transport plane and three crop dusting planes to relief efforts for ten days. India has disputed claims made by Pakistan that it blocked its military access to move helicopters from West Pakistan. Pakistan's government pledged an extra $116 million for disaster aid on November 24. Yahya Khan arrived in Dhaka on November 24 to take charge of the rescue efforts. Vice Admiral S. M. Ahsan, governor of East Pakistan, refuted claims that the military did not respond soon enough and stated help was being sent across the affected region, with the exception of a few isolated areas.

President Khan said a week after the hurricane made landfall that his administration had made "slips" and "mistakes" in its management of the recovery operations. He said that not enough people realize how bad the tragedy really is. Denying speculations that the general election scheduled for December 7 would be postponed, he also maintained that the poll will go on as planned, although with delays in eight or nine of the worst impacted areas. Due to a national strike and later a prohibition on government operations in East Pakistan by the Awami League, the Dhaka offices of the two government agencies directly engaged in relief efforts were shuttered for at least two weeks in March as the conflict between East and West Pakistan escalated. The field relief operations kept going, but the long-term planning was halted.

Criticism of Government Response

Even though we have a sizable army, the British Marines are responsible for burying our fallen. The early reaction of the central government was heavily criticized by political figures in East Pakistan. Ten days after the hurricane struck, eleven East Pakistani political figures issued a statement accusing the government of "gross carelessness, callous apathy, and total disregard." The president was also accused of trying to downplay the story. On November 19, students in Dhaka marched in protest at the government's response time, and on November 24, Maulana Abdul Hamid Khan Bhashani spoke to an assembly of 50,000 people, accusing the president of incompetence and demanding his resignation. Political opponents of the president said he botched the attempts and called for his removal.

When a disagreement erupted over the ownership of 20 rafts supplied by the British Red Cross, the Pakistan Red Crescent started to function independently of the government. Two days passed before a pesticide business was granted permission to use two crop dusters already present in the nation to drop off supplies in the impacted areas. There was just one helicopter sent by the Pakistani government to aid in relief efforts, and this was because Yahya Khan decided it was useless to send any helicopters from West Pakistan since they couldn't transport supplies.

Early in the new year, a reporter for the Pakistan Observer visited a week in the hardest impacted districts and reported that none of the tents donated by relief organizations was being utilized to shelter survivors and that subsidies for constructing new homes were inadequate. In January, the coldest month of the year in East Pakistan, the National Relief and Rehabilitation Committee, headed by the editor of Ittefaq, said thousands of survivors from the storm were "passing their days under [the] open sky." Families who lost their homes in the hurricane are being given up to 250 rupees (US$55 in 1971 currencies; $321 in 2016) to help them rebuild, according to a spokesperson, but he worries that the survivors would "eat the cash" since supplies are limited.

Political Consequences of 1970 Cyclone

A large portion of the electorate was unhappy with the national government's inability to give aid in December 1970, which helped propel Bangabandhu Shaikh Mujibur Rahman and his Awami League to a resounding win in the national elections. The storm forced the postponement of elections for nine seats in the National Assembly and eighteen seats in the Provincial Assemblies to January 18. In East Pakistan, the resistance movement gained ground as a result of the government's handling of relief operations, which infuriated the locals there. In addition to the sluggish distribution of funds, transportation delays meant that relief goods took longer to reach the affected communities. In March, when tensions rose, foreign workers left the country out of concern for their safety. The Bangladesh Liberation War broke out in March as a result of the deteriorating circumstances. In December of 1971, this war escalated into the Indo-Pakistani War, which ended with the birth of Bangladesh. This was one of the first examples of a natural disaster contributing to the outbreak of civil strife.

3.5 Economic Disparity Between East and West Pakistan: Consequences in Disaster Vulnerability

According to published sources, between 1960 and 1965, West Pakistan's GDP per capita grew at an annual rate of 4.4%, whereas East Pakistan's grew at a pace of 2.6%. East Pakistan provided a large portion of Pakistan's export profits through the sale of Bengali jute and tea, hence lawmakers from that region campaigned for more autonomy. Until the 1980s, East Pakistan was responsible for almost 70% of Pakistan's export revenues; however, as global demand for jute waned, that number steadily decreased.

By the time Bangladesh won its independence in 1971, East Pakistan was contributing less than half of the country's export revenue. The first foreign currency accounts and overseas trade offices were required by Bangabandhu Shaikh Mujib in 1966. By the 1960s, West Pakistan's quality of living had risen thanks to Ayub's "Decade of Progress" and the country's successful green revolution in wheat and the growth of markets for West Pakistani textiles, but East Pakistan's standard of life remained at an abysmally low-level. West Pakistan, where the federal government is located, got more foreign help than East Pakistan, where the majority of the population lives.

It is often believed that the disparity in economic opportunities between East and West Pakistan in the 1960s was a major factor in the rise of Shaikh Mujib's Awami League, which eventually led to the independence of Bangladesh in 1971. Over the last 40 years, this gap has widened, and in 2011, Pakistan's per capita GDP was 1.7 times that of Bangladesh, up from 1.6 in 1971. Even though it's been 40 years since the Fall of Dhaka and the establishment of Bangladesh on December 16, 1971, the event is still often discussed. To commemorate the 40th anniversary of Bangladesh's independence, the Daily Star, a Bangladeshi daily, published an article by Akbar Ali Khan on the topic. Although economic development in East Pakistan was revived under Ayub Khan's so-called decade of reforms, the growth rate in former East Pakistan was significantly lower than that of West Pakistan, as Mr. Khan explains in his Op-Ed. Mr. Khan, in his eagerness to justify independence on the basis of the economic argument, has obviously disregarded the following realities:

According to Mr. Khan's description, the gap between West and East Pakistan's per capita earnings in 1969–70 was $1.6. According to 2011 IMF figures, however, this ratio has risen to 1.7. While Pakistan is a medium-income nation and far above the list of least-developed countries, Bangladesh is still classed as a low-income and least-developed country by the World Bank. According to the World Bank, Bangladesh has the eleventh highest proportion of its people living on less than $1.25 a day. India, Pakistan's neighbor, ranks just 14th on this list of impoverished countries. All the other countries in this list can be found in and around the Sahara Desert. The 1947 partition of Pakistan left East Pakistan economically weaker than West Pakistan already was due to the exodus of its Hindu Bengali business elite. There were also many Muslim businesses who had moved to West Pakistan, especially Karachi, following the partition of India in 1947 who were not able to help.

Although East Pakistan benefited from Ayub's economic reforms in the 1960s, the fact that these benefits were perceived as a dispensation from a quasi-colonial military regime to its colony—East Pakistan—proved to be lethal, as Pakistani economist Dr. Ishrat Husain puts it. "Regional disparities lay at the heart of the crisis," he writes. "This was evident in ever-widening disparities in per capita incomes, in the East."

The Pakistani government's efforts to exert full economic control over East Pakistan included a wide range of strategies. The urban sector's infrastructure gap between the Eastern and Western wings was already wide, but it widened dramatically as a result of the rush by immigrant businesspeople to invest in the Western sector. Thus, agriculture and services accounted for 70 and 10% of East Bengal's GDP, whereas they accounted for 54% and 17% in West Pakistan. There was a clear discrepancy between East Pakistan and West Pakistan when it came to the launch and allocation of disaster-related projects, as shown by several data, including Pakistan's yearly development plan. What follows is a compilation of information obtained from many resources (Fig. 3.4; Table 3.4).

It is clear that bias toward East Pakistan in every sphere has existed from the country's inception in 1947. Most development initiatives, such as flood or cyclone protection projects, were seen by East Pakistanis as unfairly benefiting West Pakistanis due to the dominance of West Pakistani public officials in the country's central governing apparatus. Pakistan's economic and social consequences during the

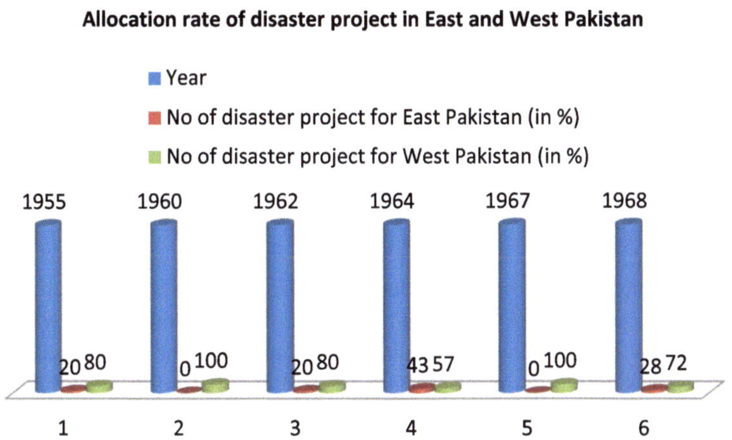

Fig. 3.4 Allocation rate of disaster project in East and West Pakistan

Table 3.4 Disaster management-related projects for East and West Pakistan from 1950–1970

Year	No of disaster project for East Pakistan (in %)	No of disaster project for West Pakistan (in %)
1955	1 (20%)	4 (80%)
1960	0 (0%)	3 (100%)
1962	2 (20%)	8 (80%)
1963	3 (44%)	4 (56%)
1965	0 (0%)	2 (100%)
1968	7 (28%)	18 (72%)

Source Information/data collected and compiled by the Researcher

last 60 years are a jumble of contradictions. Since 1947, the economy has grown by an average of 5% yearly, a pace attained by just a handful of other nations. The confluence of religious fanaticism, sectarianism, ethnic cleavages, and regional economic inequalities, however, has rendered the country politically volatile and dangerous. When compared to Pakistan, several East Asian nations that lagged behind in the 1960s have now made tremendous economic and social gains. As a result, Pakistan has not been able to fully develop.

Since a market economy cannot function effectively in an environment of political uncertainty and volatility, action was clearly required. The answer was to invest more resources into the military, which is, by far, the most professionally run organization in the nation. The Pakistani armed forces were widely recognized as the country's de facto border guards, thanks to their merit-based entrance and promotion system, as well as their exceptional professional training and behavior. It was certain that it was operating in the nation's best interest and that it understood what changes were necessary to jumpstart the economy. Prior military dictators always toppled democratically elected governments on the grounds that they were bad for the economy. Military regimes often assumed they were the only ones able to deliver the market-enabling attributes of openness, stability, consistency, and predictability.

3.6 Disaster Preparedness During British India and East Pakistan

According to the available literature, during the British era, field-level authorities, in particular those responsible for tax and revenue collection, had to deal with natural catastrophes by offering some humanitarian assistance to the direct victims. They have done work like this on top of their usual duties before. Every officer in the District now has the responsibility of overseeing them. There were no specialized or independent organizations for responding to disasters that could be located in the research. Even yet, there was no early warning system in place at the time. There was no proof of preventative action in the catastrophe zones. When calamity struck India, the British administration never bothered to provide aid and instead displayed a cold and uncaring attitude toward the victims.

Flooding in Pakistan's first yet devastating natural catastrophe occurred in 1950, affecting more than 10,000 communities and killing 2,910. Since then, East and West Pakistan have been plagued by frequent natural disasters such as floods, droughts, cyclones, earthquakes, and landslides, but no administration in the region, including Pakistan's, was able to come up with a plan for dealing with them until Bangladesh gained independence.

Only in the late 1960s did East Pakistan (now Bangladesh) recognize it needed a flood control program, which led to the program's inclusion in the country's yearly development plan in 1954 and 1955. Despite this, however, efforts in this area were minimal. In 1955, they continued to include emergency response activities for natural disasters in their yearly growth plan. The Emergency Relief Cell (ERC) of the Cabinet Division is the federal government's point of contact during crises in both East and West Pakistan, as outlined in the 1955 annual development plan. The ERC works with provincial relief agencies and relief commissioners to ensure that disaster supplies are distributed efficiently throughout the country [27] (Fig. 3.5).

Responsibilities of the ERC in connection with disaster relief were:

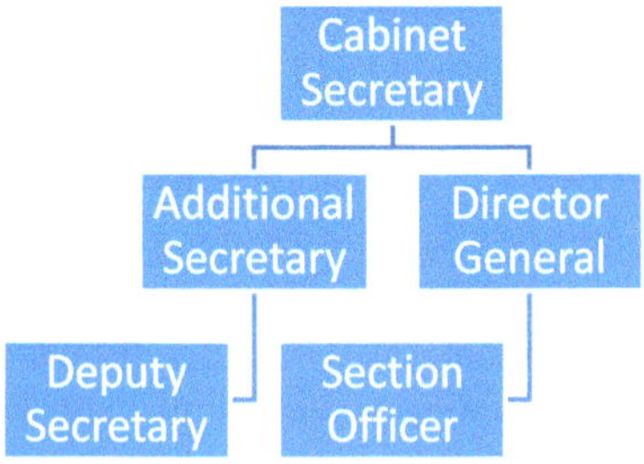

Fig. 3.5 Organogram of ERC, developed by the researcher

A. To provide in cash as well as in kind to supplement the resources of the provincial governments in the event of major disasters.
B. To coordinate the activities of the federal Division, Provincial Governments, as well as governmental, semi-governmental, international and national aid-giving agencies, in the conduct of operations for relief of disasters.
C. To maintain contact with international aid-giving agencies/ voluntary organizations and donor countries for disaster relief measures.
D. To administer relief funds, being maintained at the Federal Level.
E. To stockpile certain items of basic necessity and establish central inventory of resources.
F. To provide assistance to the calamity stricken friendly countries.
G. The ERC operates an Emergency Control Room, which coordinates the situation during calamities by liaising with relevant agencies such as the Federal Flood Commission, Meteorological Department, and Provincial Governments.
H. The ERC maintains a warehouse in the capital, Islamabad, stocking essential non-perishable relief items such as medicines, blankets, clothing, and tents. In addition, there is a Relief Goods Dispatch Organization (GDO) located in Karachi. This is responsible for receiving and dispatching all relief goods from foreign and local agencies in the event of a disaster.

East Pakistan Government in 1955 proposed the following Cyclone and Flood Management Programme:

Structural measures	Non-structural measure
a. Construction of embankments b. Construction of dikes/gabion walls/flood walls c. Channelization of flood waters d. Construction of delay action dams	a. Improved flood forecasting system through b. Effective data collection and dissemination system c. Real-time rainfall and river flow data collection d. Weather radar prediction e. Modern system of transmission of flood forecasts

A scientific and service division, the Met Department began operations in the early 1960s. It was in charge of the weather forecasting for both East and West Pakistan. The Meteorological and Geophysical Department's primary roles included forecasting, predicting, and modifying weather conditions, as well as assisting with disaster preparedness related to weather and geophysical phenomena and the development of agriculture in accordance with the country's climatic potential.

Provincial Relief Department was established in 1963 with a view to.

A. Providing adequate resource support to area administration through coordination with Provincial Government Departments/Agencies.
B. Keeping provision of necessary funds to the area administration for relief work.
C. Overseeing the working of area administration for relief work.

D. Obtaining field reports of losses and apprising the Provincial Government/Federal Government.
E. Assessing and evaluating losses and suggesting to the Federal/Provincial Governments for providing relief to the affected persons.

3.7 Indigenous Knowledge and Coping Strategy of the Disaster Vulnerable Community

We all know that people's perceptions play a significant role in shaping their coping strategies. Community members' subjective definitions of risk circumstances shape how individuals and society respond to threats. What constitutes an undesirable state of affairs or negative outcome is subjective and depends on the individual's values, interests, and level of expertise. Indigenous knowledge has developed through thousands of years of observation, experimentation, and careful observation of the natural world by humans. East Pakistan and British India might learn a lot from their indigenous people's understanding on how to deal with and recover from natural calamities. Preparedness and reaction to disasters are topics we'll be covering here from the perspective of their indigenous knowledge systems.

According to the literature, throughout British India and East Pakistan, village elders used their traditional knowledge system to foresee meteorological conditions, identify the kind of flood or cyclone they may expect, and take precautions to lessen the impact of such events. In many instances, their forecasts were almost identical to the forecasting message put out by the relevant governments at the time [28]. Not much of a difference existed in the conventional knowledge system across socioeconomic classes. Predictions of early wages are therefore identical for all categories of marginalized men and women (the homeless, landowners, women who lead homes, individuals with disabilities, members of occupational minorities, ethnic minorities, fisherwomen, and members of religious minorities such as Hindus). The symptoms of heavy rain, flooding, flash flooding, and riverbank erosion as described by numerous cultures have been compiled. The symptoms and signs are described in more detail below.

When the outdoor temperature is high throughout the months of Jyaishtha to Ashwin, the locals see this as a signal that rain and flooding are on the way. They predict that within a few days, flash flooding will occur due to the high temperatures and subsequent thick clouds. There is widespread use of this ancient foreknowledge among residents. Most adults over the age of 65 have been shown to depend rather strongly on this information.

You may also use the hue of the clouds as a reliable indicator of impending rain. It is believed that black or dark ash-colored clouds would bring torrential downpours. If a massive, ominous cloud stays put over the horizon for many days, it's likely that the atmosphere will be gloomy and wet. In the next 8–10 days, flash floods are common due to this sort of precipitation.

It's also thought that if the border's mountainous region experiences heavy rains for two or three days straight, flash flooding would likely occur in the low-lying districts nearby. As the water level in the surrounding rivers rises due to this severe and constant rainfall, the lands quickly get inundated. The river's water hue is often employed as a cyclone and flood indicator. Researchers have shown that a change in the clarity and darkness of river water is a reliable indicator of impending flood or storm, as well as riverbank erosion, in the regions around the river being monitored.

It's important to remember that certain animals' habits may serve as an early warning system for flooding or cyclones. When ants emerge from the ground with their provisions and eggs, it is a sign that rain is on the way, according to local superstition. In addition to ants, frogs' behavior may give information that can be used to foretell flooding or cyclones.

Predictions on the character of the rainy season and powerful storm are often made using the "water in arum leaves" approach. For this purpose, we have chosen to test certain arum plants with leaves. Twelve leaves and their stems are harvested from the plants that were chosen. Following this, the Bengali names for each of the chosen leaves are revealed. So, each leaf will stand in for a different month [29]. Ropes are used to bind the chosen leaves together. The next morning, someone is tasked with reopening the fastening of the leaves so that the water collected in each leaf may be measured. The community always chooses an individual or group of individuals to take on this responsibility.

How Did They Cope with Disaster Shocks Are Briefly Put Below?

The term "social capital" encompasses the set of institutions, social ties, and shared beliefs that determine the extent and nature of social connections within a community. Capital is a key concept, and these complex webs of relationships may be used to boost people's prospects in today's economies. Social capital seems to be a robust type of assets for many people during times of flood or hurricane. Each person has their own unique stock of social capital, and it takes many forms. It's true that having access to these resources helps people deal with tough times in their lives. Close family bonds flourish and deepen in the wake of natural disasters like floods, cyclones, and earthquakes. During times of disaster, such as a flood, it is often the wealthy relatives who assist out those in need. Relatively well-off relatives give aid in the form of food, temporary housing, and loans with favorable conditions, but they also expect their dependents to help out around the house in exchange.

As a result of the persistent flooding, locals have resorted to "rising plinth" techniques as a means of safeguarding their homes. As a result of the high population density in the char, a term for the low-lying flood plain region, most of the homes in the area are situated on higher ground. From the char land bed heights, most homesteads are 5–6 feet. These types of homes are worth replicating in other areas where vulnerable individuals become powerless in the face of danger since the approach is simple and safer for emergent periods for refuge. People in the area often fortify their homes with sturdy pillars and scaffolding to protect them against natural disasters like floods, water surges, and river erosion [30]. This is a frequent way for individuals to be ready for emergencies. Both sexes were present at these DIY home repair events.

Because many locals rely so heavily on revenue from domestic animals, their pets and livestock often take priority above their own safety. As a result, they came up with a novel solution to this problem, which has far-reaching effects on their economic and social lives: they constructed a makeshift scaffolding system around a designated area of the farm. One method of safeguarding domestic animals and family belongings from floods, water surges, and other calamity occurrences is to construct a scaffold by elevating some area on the farm three feet high. All the household belongings, including the dogs, are moved to the higher scaffolding when there is a flood. After the flood waters subside, the land is used as a plateau for planting a new, very valuable crop of vegetables, such as gourds, pumpkins, beans, and other cucurbitaceous plants.

Food insecurity is a major issue in areas affected by disasters and in the aftermath. Scaffold was used to store grains for future usage. In addition to storing food for domestic animals, they dried fish, chira, muri, etc. Long-term seed storage creates a diversity of plant and variety in the seeds. If farmers have access to new varieties of seeds, they may test the seeds' yield on plots of land as small as 20 or 25 decimals. If the results are positive, the farmers will save the seeds to use for next year's planting. How many acres or plots of land will be used to plant that specific seed type in the next growing season? That's the number you'll want to keep in mind when deciding where to store your seeds. While still out in the field, growers choose which plants to harvest for seed.

High-quality plants and fruits are often chosen. Farmers then use field tests to confirm their children have reached adulthood. They are then dried after being sliced and bundled individually. They separate out any unusual varieties of paddy before thrashing. The seeds are then carefully dried in the sun for two or three days in a row, for five or six hours every day, until only the completely developed seeds remain. Women sometimes attempt to dry fish in the sun in order to keep it for later when the price of fish drops or when they have any leftover. The locals call the bamboo scaffold used for drying fish "chang" (pronounced "chay"). This "chang" is usually cooked up in the kitchen or another open area of the home where there aren't any nearby trees. The fish will dry more rapidly in the sun if you do this. The "chang" has been replaced with the more comfortable "chatai" (mat).

Old nets are often rigged around and above these scaffolds to protect the dried fish from crows and other birds. Because so many tube wells are submerged, people are deprived of a safe source of drinking water during a flood. As a result, homeowners are re-excavating the ponds in front of their homes and elevating the pond's perimeter in an effort to better safeguard the water supply from potential threats like floods and water surges. Most farmers keep cows as pets because they provide a reliable source of revenue in times of financial hardship, and the cow dung they collect serves as both fuel and manure for their farms. Thus, the preservation of cow dung is an essential concern for those who live in flood- and cyclone-prone regions and rely on it as a fuel source. Although the respondents knew that cow dung had long been used in the region as a culinary ingredient and fertilizer, they were unable to pinpoint precisely when the practice had first begun.

3.8 Community-Based Disaster Management Approach

Top-down management's shortcomings have been exposed, and it's now clear that vulnerability reduction is essential for successful disaster management. This strategy has failed to meet the requirements of marginalized groups. It becomes clearer as we learn more about catastrophes and losses that the exponential rise in the frequency of small and medium-sized disasters is mostly responsible for the rise in the severity of disaster-related losses. Therefore, many experts and stakeholders believe it is crucial to shift to a new approach that actively includes vulnerable people in the process of preparing for and responding to disasters, as well as in the subsequent efforts to recover. The widespread support for this bottom-up strategy stems from the fact that it recognizes communities as the most reliable arbiters of their own safety and as competent to make the most appropriate choices for their own well-being.

In an effort to find a more modern method, the concept of Community-Based Disaster Management (CBDM) was developed. CBDM's goal is to increase resilience to hazards and catastrophes by lowering the number of individuals who are vulnerable to them. To effectively mitigate catastrophe risks, it is necessary to first conduct a comprehensive examination of a community's vulnerability and capacity, as well as its exposure to hazards. When a community is engaged from the beginning, its needs and resources, both current and future, are taken into account. As a result, there is a better chance that solutions will be implemented. Rather than only focusing on the mechanics, the public is encouraged to provide input on the substance as well. Improved disaster risk management is expected to have immediate benefits for the local community.

3.9 Volunteerism in Disaster Management

While the vast majority of Bangladesh's volunteers operate on an ad hoc basis, a few notable formal volunteer organizations are also active in the country, providing early warning, emergency relief, and recovery support to the government of Bangladesh and other actors involved in disaster management.

Since its inception in 1973, the Bangladesh Red Cross Society (BDRCS) has worked to implement the guiding principles of the international Red Cross and Red Crescent movement and alleviate the suffering of those most in need as a result of natural disasters and other health crises. With a presence in every district and even every subdistrict, the Red Crescent society is integral to the government's disaster management efforts in Bangladesh. Many BDRCS programs are focused on disaster management and the betterment of disadvantaged populations.

As an early warning system, the Cyclone Preparedness Programme has been very successful. It was founded following the catastrophic typhoon Bhola of 1970, which killed over 300,000. In 13 cyclone-prone coastal areas, the CPP has installed early warning facilities to alert the public of oncoming danger. Trust in early warning information from the government of Bangladesh has been bolstered in part because of this organization's volunteer network, which helps distribute cyclone warnings to the local communities. Volunteers disseminate the data by visiting each community and discussing the impending cyclone and its potential impact.

Better coordination among stakeholders and the building of confidence between the local community and national early warning systems are made possible via the efforts of this volunteer network that acts as a bridge between the national government and local communities. The CPP has been successful because it is easy to implement, takes into account sociocultural factors, relies on community-based participation, and has an efficient early warning communications network. About 50,000 active volunteers make up the CPP today, and these groups work in tiny teams ready to distribute early warning information down to the lowest administrative levels.

The decrease in fatalities caused by major cyclones in Bangladesh over the past 40 years is the best indicator of the Cyclone Preparedness Centre's contribution to disaster management activities in Bangladesh. Though the CPP is not solely responsible for the reduction in fatalities, the program's prominent role in distributing data gathered by other government agencies and alerting residents of high-risk locations is clear. Death tolls have been progressively declining since the horrific cyclone Bhola in 1970, which took over 300,000 lives, with the most recent being the 113 deaths lost during Cyclone Aila in 2009.

References

1. Yusuf H (1999) Pakistan: a study of political developments 1947–97. Sange Meel Publications, Lahore
2. Keith C (1968) Pakistan: a political study. Allen & Unwin Ltd., Oxford
3. Allen M (1996) The destruction of Pakistan's democracy. Oxford University Press, New York
4. Administration B (1895) Bengal famine code. Bengal Administration, Calcutta
5. Administration B (1944) Bengal Famine and its impact. Bengal Administration Relief Service, Calcutta
6. Administration B (1945) Bengal Famine and British Raj response. Bengal Administration Relief Service, Calcutta
7. Administration B (1944) Bengal land revenue commission report. Land Revenue Service, Calcutta
8. Braund C (1947) Natural calamities. Asia Publishing House, Bombay
9. Bayly C, Harper T (1944) Forgotten armies: Britain's Asian empire & the war with Japan. Penguin Book, London
10. Bayly C (1945) Britain's Asian empire and development in Bengal administration. Penguin Books, London
11. Administration B (1944) Development village in Bengal. Land Revenue Service, Calcutta
12. New York Times (1970) Thousands of people in East Pakistan are killed by Tidal Wave. New York Times, New York

13. Braund C (1945) Natural calamities East Pakistan. Asia Publishing House, Bombay
14. New York Times (1943) Famine–An ignorance for poorest. New Work Times, New York
15. Schwerdt R (1949) Weather pattern and disaster in South Asia. National Oceanic and Atmospheric Administration, New York
16. Joint Typhoon Warning Center (1969) Western North Pacific tropical storms 1969. JTWC, Tokyo
17. Frank N, Husain SA (1971) The deadliest tropical cyclone in history. American Metrological Society, USA
18. Choudhury AM (1987) Monsoon flood boundary delineation and damage assessment using space borne imaging Radar and Landsat data in photogrammetric engineering and remote sensing. UN HQ, New York
19. New York Times (1970) Pakistan death toll 55,000; may rise to 300,000. New York Times, New York
20. New York Times (1970) East Pakistan–Devastation in lives and livelihood. New York Times, New York
21. New York Times (1970) Cyclone 1970: crying need of life saving support. New York Times, New York
22. The Daily Observer (1970) Living with death: save us. The Daily Observer, Dhaka
23. USAID (1971) USA in East Pakistan cyclone 1970. USAID HQ, USA
24. Government of Canada (1970) Relief assistance to Pakistan cyclone 1970. Canadian High Commission to India, Delhi
25. UNO (1970) Appeal with assistance for Pakistan. UN HQ, New York
26. Government of West German (1970) Relief for lives. Government of West German, Berlin
27. East Pakistan Government (1966) Crisis recovery programme. Planning Commission, Dhaka
28. ADPC (2000) Mainstreaming disaster risk reduction into development policy. Planning and Implementation in Asia, Bangkok
29. ADB (2004) Disaster coping method in South Asia. Asian Development Bank, Manila
30. Government of East Pakistan (1956) Annual development programme. Planning Commission, Dhaka

Chapter 4
Legal and Institutional Advancement of DM in Bangladesh

4.1 Introduction

Resilient communities are built on a foundation of laws and rules. They are crucial for mitigating the effects of natural disasters, averting further harm, and enhancing public security. The Hyogo Framework for Action from 2005 emphasized the need for effective laws to aid in the mitigation of disasters (DRR). Since its adoption in March 2015, the Sendai Framework for disaster risk reduction has emphasized the need of reevaluating and strengthening existing legal frameworks. A national DRR system is established by legislation, which sets priorities, institutional mandates, and other elements.

DRM laws vary in how much they address issues like early warning systems, community and civil society engagement, resource allocation, and education and public awareness. Depending on the jurisdiction, these concerns may be dealt with only by the DRR legislation, or by additional or supplementary laws that are also part of the legal framework. To further incorporate disaster risk reduction into government development planning, an annual work plan for the Ministry of Disaster Management and Relief is developed with the help of the parliamentary standing committee on the Ministry.

Bangladesh is more at risk from several natural and climate-induced catastrophes due to its location and low-lying features. In terms of population, it is the third most susceptible nation to sea level rise, and it is in the top ten for the proportion of its population that lives in low-lying coastal zones. Since its independence, Bangladesh has been proactive in formulating a wide range of policy frameworks, including the Standing Order on Disaster and the Bangladesh National Building Code. However, the country's track record of successfully implementing these policies has been less than stellar in many cases. Although the DM Act was initially conceived of in Bangladesh in the mid-1990s, it took the other South Asian nations over 20 years to adopt similar legislation [1]. Following Bangladesh's lead, nations like India and

M. T. Islam, *Disaster, Governance and Development*, Disaster Studies and Management,
https://doi.org/10.1007/978-981-99-1631-3_4

Sri Lanka approved the DM Act and SOD relatively early on. The government of Bangladesh has developed a basic model to aid in disaster preparedness and response [2].

4.2 Disaster Management Model in Bangladesh

The model's three pillars guarantee that advancing toward a more all-encompassing culture of risk reduction will continue to be prioritized (Fig. 4.1).

This component of the model advocates for using scientific analysis (including the effects of climate change) as the foundation for precisely predicting the future

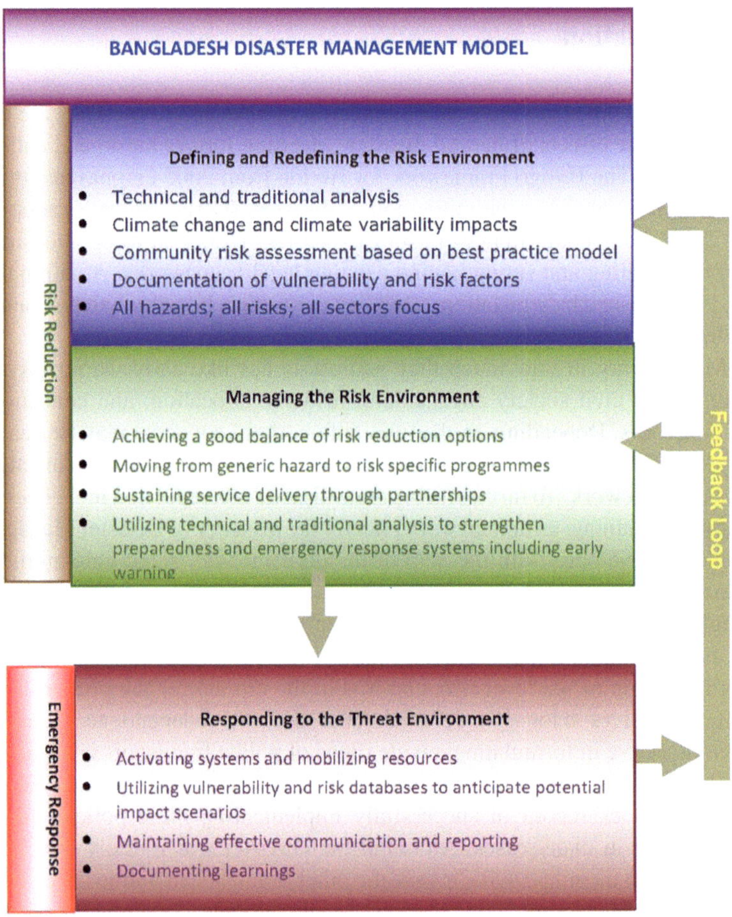

Fig. 4.1 Disaster management model in Bangladesh. *Source* National Disaster Management Plan, 2008 [3]

risk environment with regard to all risks, all sectors, and all geographic locations. The risk environment has to be characterized before any measures can be taken to manage it (risk treatment alternatives can be developed). By following a systematic and analytical procedure, defining the risk environment leads to an understanding of the relationship between the hazards and the vulnerable components (the community) [3]. Traditional hazard analysis and formal hazard analysis both play a role in defining the risk environment, which entails the following processes [4]:

- Understanding the social, political, and community environment (Establishing the context).
- Establishing what are the likely threats (Identifying hazards and risks).
- Understanding the likelihood and consequences (Analyze the risks).
- Rank risks in priority (Evaluate risks).
- What can be done to eliminate, reduce, or manage risk (Identify risk treatment strategies).

Hazard Analysis

Risk analysis is the study of potential sources of injury or loss. It may be done informally or formally, and it has to take into account both fundamental risks (such a cyclone) and secondary hazards (e.g., storm surge, wind, rain). Details on the impact that a hazard will have on communities should be included in any representation of a hazard that makes use of a geographic information system (GIS) or other current mapping methodology [5].

Vulnerability Assessment

Vulnerability assessments are essential for learning about the cumulative effects of hazards on populations. This should be done taking into account community members (such as women, children, and the impoverished), community infrastructure (such as energy, transportation, and community services), and community livelihoods (e.g., food, accommodation, farm activity, industry) [6].

Risk Treatment

Prioritizing risks and identifying strategies to mitigate or eliminate them are necessary steps in any risk treatment plan.

Managing the Risk Environment

As a result of the risk assessment, this component of the approach encourages the construction of risk reduction initiatives (Community-Based Adaptation Programmes). Programs for hazard prevention, preparedness, response, and recovery may then shift from being general in nature to being risk-specific. As a result, communities will be able to adapt to a changing risk environment by taking preventative measures [7]. Creating plans and methods to do away with or lessen the risk is a key part of risk management.

The term "mitigation" is used to refer only to technical solutions, but nowadays it's understood to include everything done to get rid of or lessen risk (e.g., community education and awareness, planning activities, development of warning systems). Activities in these four categories (prevention, preparedness, response, and recovery) were formerly referred to as the PPRR Model. Legally acceptable mitigating actions include the design and implementation of response and recovery systems.

Responding to the Threat Environment

This part of the concept entails doing something in response to a real-life dangerous circumstance. It aids Bangladeshi disaster management professionals in explaining the distinction between preventing disasters and responding to them, as well as how a more precise understanding of risk contexts may improve the effectiveness of both. Risks and hazards are not always able to be controlled. There may be instances when you have to react to a new danger or an already-existing one. There is a requirement to engage the risk management reaction and recovery mechanisms in this circumstance. Possible reactions are [8]:

- Warning Period (Alert and activation),
- Hazard Onset (Response), and
- Post-Hazard Period (Relief, early recovery, and rehabilitation).

The key attributes of the model are:

- It provides a framework to guide the achievement of the Hyogo Framework for Action commitments.
- It clearly articulates the key elements of disaster management and their interactive relationships.
- It facilitates the transition from generic hazard-based to specific risk-based programmes through the inclusion of technical inputs.
- It provides guidance for the design of policy, planning, and training.
- It provides a mechanism to achieve consistency in process and methodology.
- It ensures preparedness and response strategies are influenced by technical and traditional considerations.

Mainstreaming Risk Reduction—The Strategies

For all hazardous risk reduction programs throughout the nation to be sustainable, they must be embraced by government, non-governmental organizations (NGOs), and the business sector. In Bangladesh mainstreaming is considered in much the same light as poverty reduction in that it is the product of several top-down and bottom-up actions.

Advocacy: Building knowledge and understanding of the benefits of risk reduction and the roles these organizations play in implementing risk reduction programmes requires a concerted effort to increase awareness among Political, Senior Policy and Government Department Officials, Media and Academic Institutions.

Policy and Planning Reform: To better enable mainstreaming and create a complete risk reduction culture, the policies governing disaster management and development planning are now undergoing a thorough revision.

Capacity Building: The purpose of this plan is to ensure that disaster management committees (DMCs) at all levels accurately represent risk reduction and emergency response duties by conducting a comprehensive evaluation of their roles and responsibilities. To guarantee that committees obtain capacity development training and can perform their duties competently, a nationwide training program is now being designed.

Planning Frameworks: DMC plans are being revised extensively to include risk reduction mainstreaming at all tiers of planning for disasters.

Uniform CRA Guidelines: Consistent community risk assessment (CRA) methodologies that are also compatible with the risk reduction planning processes of the various DMCs are now being implemented. Steps are included in the recommendations to guarantee solid connections with data from scientific analyses (Fig. 4.2).

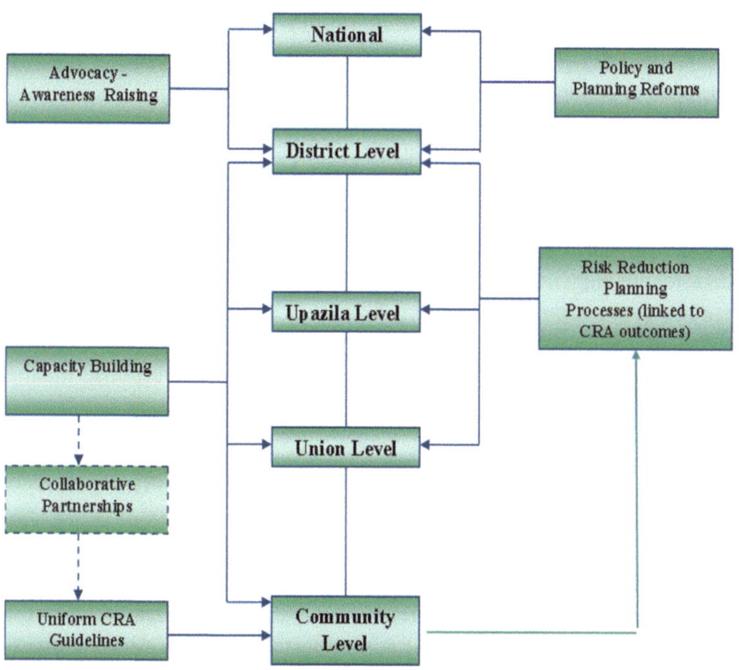

Fig. 4.2 Different stages of disaster management system in Bangladesh. *Source* Haque et al. (1992) [14]

Priority Areas of DRR Legislative Affairs in Bangladesh

- Focuses on emergency response to natural hazard situations; however, it may also contain components of immediate preparation, early warning, and recovery.
- Includes the core DRR activities of hazard avoidance, preparation, reduction, response, and rehabilitation. Elements of DRR are included, but neither cross-sectoral procedures for DRR nor regulation of a variety of linked sectors, such as DRR funding, risk mapping, early warning, or particular mechanisms for DRR education, are often included.
- Identical topics to the Broad DRR statute, but DRR is given unambiguous priority, which might be seen as allowing for a "whole-of-society" approach. Besides the primary DRR activities, it often addresses a wide range of linked topics and the corresponding local institutional structures and/or duties.
- Clearly prioritizes DRR, but does not provide a comprehensive and detailed range of related subject matter because this is covered by a number of other laws that may include, but are not limited to, laws on specific hazards, laws on natural resource management, building and construction laws, and local governance laws.

Standing Orders on Disaster (SOD) have been greatly enhanced in the current iteration. The new edition includes, among other things: (1) a more streamlined table of contents, (2) a more thorough list of definitions, (3) a list of abbreviations, (4) a summary of the regulatory framework for disaster management, (5) a new section on multi-agency disaster incident management, and (6) a description of the roles and responsibilities of all parties involved in reducing risk. As a result, not only in times of emergency but also in everyday life, it must be adhered to [9]. The Standing Orders on Disaster (SOD) are the basic backbone of the country's national disaster management system, outlining the various departments' duties and how they should work together to better prepare for and respond to disasters.

A disaster management regulatory framework is constructed under which the Bangladesh Disaster Management Framework is implemented and in which the activity of Ministries, Departments, NGOs, and civil society is carried out. Bangladesh's disaster risk reduction (DRR) and emergency response management (ERM) activities are managed and executed within a legal, policy, and best-practice framework provided by the disaster management framework. The skeleton consists of:

1. Disaster Management Act.
2. National Disaster Management Policy (to be formulated and adopted).
3. National Plan for Disaster Management.
4. Standing Orders on Disaster (SOD).
5. Guidelines for Government at all Levels (Best Practice Models).

To aid ministries, NGOs, disaster management committees, and the general public in carrying out disaster risk management, best-practice models in the form of governmental guidelines are established. Among the many things covered by the rules are:

1. Community Risk Assessment (CRA) Guideline
2. Disaster Impact, Damage, Loss, and Need Assessment Guideline
3. Local Disaster Risk Reduction and Emergency Fund Management Guidelines
4. Indigenous Coping Mechanism Guidebook
5. Hazard Specific Risk Assessment Guidelines for Earthquake, Flood, Cyclone, Fire, Drought, etc.
6. Emergency Response and Information Management Guideline
7. Contingency Planning Templatc
8. Emergency Response Management Guideline
9. Disaster Information Management Guideline
10. Sectoral Disaster Risk Reduction Planning Template
11. Natural Disaster Risk Incorporated Local Level Planning Template
12. Guidelines for Road, Water, Industrial and Fire Safety
13. Guideline for Disaster Shelter Management
14. Monitoring and Evaluation Guideline for the Implementation of the Plan
15. Guideline for International Assistance Management in Disaster Emergency
16. Multi-Agency Disaster Incident Management Guideline.

4.3 Development of Disaster Management Institutions in Bangladesh

Department of Disaster Management

Following the passing of the Disaster Management Act, a Department of Disaster Management (DDM) will be established inside the Ministry of Disaster Management and Relief. The Department will be responsible for carrying out the goals of the Disaster Management Act by, among other things, conducting humanitarian assistance programs efficiently to increase the capacity of the poor and the disadvantaged, strengthening, and coordinating programs undertaken by various government and non-government organizations related to disaster risk reduction, and coordinating emergency response efforts. In the event of a catastrophe, the government issues directives and suggestions, and it is the responsibility of DDM to carry out these policies and the national disaster management plans.

Directed by the Director General, DDM coordinates efforts between government agencies to better prepare for and respond to natural disasters. These agencies include those responsible for science, technology, research, and education; development partners; United Nations agencies; and non-governmental organizations. DDM serves as a think tank for the relevant Ministry of the Government of Bangladesh by conducting research, hosting workshops and training programs, publishing reports and papers, and offering policy advice services. DDM's long-term goal is to become an internationally recognized hub for disaster management education, research, and capacity building at all levels of the field.

Mission and Vision

The Department of Disaster Management (DDM) will be established when its Vision, Mission, Function, and Modalities have been defined in conjunction with the Ministry of Disaster Management and Relief and in light of the Disaster Management Act.

Vision

The Department of Disaster Management (DDM) aspires to be a thriving center of expertise in the areas of disaster risk reduction (DRR) mainstreaming into the Disaster Management Programme; reducing the vulnerability of peoples, especially the poor and disadvantaged, to the various impacts of disasters; and knowledge, research, and capacity building on the entire disaster management cycle in accordance with the DM Act.

Mission

By conducting risk reduction activities, efficiently responding to disaster events, and strengthening and coordinating programs undertaken by different stakeholders related to DDR and DRM, the Department of Disaster Management (DDM) would serve the Ministry of Disaster Management and Relief in its efforts to implement the objectives of the Disaster Management Act.

Functions and Responsibility of DDM

- To reduce the overall vulnerability from different impacts of disasters by undertaking risk reduction activities.
- To conduct different activities to efficiently boost the moral and the rehabilitation programmes.
- To strengthen and coordinate programmes undertaken by various government and non-government organizations related to disaster risk reduction and emergency response.
- To execute the directions and recommendations by the government in connection to disaster management.
- To execute the national disaster management principles and national disaster management planning.
- To undertake any programme that is deemed fit to support the effective disaster management infrastructure for all kinds of disaster.

Directorate of Relief and Rehabilitation

The Department of Relief and Rehabilitation (DRR) and the Disaster Management Bureau (DMB) will be consolidated into the new Department of Disaster Management (DDM) under the terms of the proposed Disaster Management Act (DDM). The Department of Relief and Rehabilitation (DRR) is responsible for a wide variety of crucial tasks, such as aiding the Ministry of Food and Disaster Management in the creation and implementation of programs and policies, and overseeing and carrying out all food-related aid programs like the Food for Work (FFW) and Test Relief (Test

Relief) schemes and relief operations. DRR is also heavily involved in the distribution of house-building funds, the implementation of rehabilitation programs, the Vulnerable Group Feeding (VGF), the provision of gratuitous assistance, the sanction of the carrying cost of food grains of FFW, and other similar activities. Also, DRR helps with things like building infrastructure like minor bridges, culverts, food, and cyclone shelters, and preparing lists of beneficiaries when natural disasters strike.

Disaster Management Bureau (DMB)

Under the direction of the high-level Inter-Ministerial Disaster Management Coordination Committee, DMB is a small, dynamic professional unit at the national level that provides specialist support functions in close collaboration with district- and Upazila-level authorities and the concerned line ministries (IMDMCC). The Disaster Management and Relief Division (DMRD) relies on this technical arm to coordinate disaster relief efforts at all levels, from the federal to the local. The DMB Development Goals are as follows:

- Increase awareness at all levels of society to reduce disaster risks and losses.
- Strengthen national institutional capacity for disaster management up to district, Upazlia, and union levels.
- Establish disaster action plans in the most disaster-prone districts and Upazilas.
- Enhance the knowledge and skills of key personnel having disaster management responsibilities.
- Promote proven local-level risk reduction ("proofing") measures.
- Improve the effectiveness of warnings and warning dissemination systems.

An all-risk management approach is being used to design relief and recovery efforts, with the goal of bolstering the resilience of vulnerable populations and reducing their exposure to individual risks. The Ministry of Relief and Rehabilitation was renamed the Ministry of Disaster Management and Relief in 1999, and again to the Ministry of Food and Disaster Management in 2003 to reflect the shift in focus from relief and response to Comprehensive Disaster Management (MoFDM). To guarantee that catastrophe episodes were planned and coordinated in conformity with the Standing Order on Disasters (SoD), a system of interconnected entities was established [10].

The Disaster Management Bureau (DMB) was established in 1992 as a national-level professional entity under the Ministry of Disaster Management and Relief as part of the aforementioned paradigm change. DMB was tasked with providing specialized support functions, and it does so in close conjunction with District and Thana/Upazila level; authorities and the appropriate line ministries, all of which fall under the supervision of a high-level Inter-Ministerial Committee (IMDMCC). DMB is a high-level Inter-Ministerial Committee that oversees a small, active professional unit at the national level that provides expert support services in close coordination with District and Thana-level authorities and the relevant line ministries (IMDMCC). The Ministry of Food and Disaster Management (MoFD) has a technical branch responsible for coordinating and overseeing disaster relief efforts at all levels, from the national to the local.

Disaster Management and Relief Division/Ministry: Development

All national disaster management operations in Bangladesh must be coordinated by the Disaster Management and Relief Division (DM&RD), Ministry of Foreign Disaster Management (MoFDM). The Standing Orders on Disaster (SOD) were published in January 1997 by the Ministry to regulate and oversee disaster management operations in Bangladesh. The SOD's stated purpose is to ensure that all relevant parties recognize and carry out their obligations in relation to disaster management at all levels. In order to carry out their duties effectively, all Ministries, Divisions/Departments, and Agencies must create their own Action Plans in accordance with the Standing Orders.

When a crisis strikes, the National Disaster Management Council (NDMC) and the Inter-Ministerial Disaster Management Coordination Committee (IMDMCC) will go to work making sure everything runs smoothly. The District Disaster Management Committee, Thana Disaster Management Committee, and Union Disaster Management Committee will each coordinate efforts at their respective levels. The DMB is there to help in whatever way it can during this time of need. To promote efficient planning and coordination of disaster risk reduction and emergency response management, a number of interconnected organizations have been established at both the national and subnational levels (Fig. 4.3).

At the national level

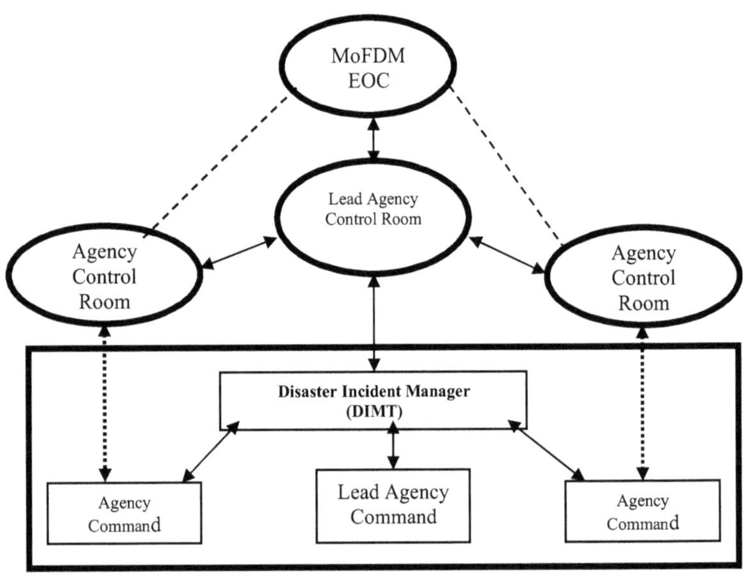

Fig. 4.3 Coordinated disaster management in Bangladesh

1. The Prime Minister presides over the National Disaster Management Council (NDMC), whose job is to create and evaluate disaster management policies and provide instructions to all parties.
2. The Disaster Management and Relief Division (DM&RD) operates under the direction of the Inter-Ministerial Disaster Management Coordination Committee (IMDMCC), which is responsible for carrying out the National Disaster Management Council's (NDMC) and the government's disaster management policies and decisions.
3. The Honorable Prime Minister has appointed a seasoned individual to lead the National Disaster Management Advisory Committee (NDMAC).
4. The Secretary of DM&RD and the Director General of DMB serve as the member secretary of the National Platform for Disaster Risk Reduction (NPDRR). For all involved parties, this hub will serve as a central point of coordination and access to vital resources.
5. The Honorable Minister for the MoFDM and the Director General of the DMB serve as Chair and Secretary, respectively, of the Earthquake Preparedness and Awareness Committee (EPAC).
6. The Secretary of the Disaster Management and Relief Division chairs the Cyclone Preparedness Program Implementation Board (CPPIB), which reviews preparatory measures taken in the early stages of a cyclone's approach.
7. The Honorable Minister, MoFDM, and the Secretary, DM&RD, serve as chair and secretary, respectively, of the Cyclone Preparedness Programme (CPP) Policy Committee. The Director General of the Disaster Management Bureau (DMB) heads the Disaster Management Training and Public Awareness Building Task Force (DMTATF), whose mission is to coordinate the training and public awareness activities of the government, NGOs, and other organizations in the event of a disaster.
8. It is the responsibility of the Director General of DMB to oversee the Focal Point Operation Coordination Group of Disaster Management (FPOCG), which is responsible for reviewing the Contingency Plan prepared by the relevant departments and coordinating the efforts of other government agencies involved in disaster management.
9. The Director General of DMB presides over the Disaster Management NGO Coordination Committee (NGOCC), which reviews and coordinates the work of relevant NGOs around the country.
10. The Director General of the DMB appoints a Committee for Rapid Dissemination of Disaster-Related Warning/Signals (CSDDWS) to investigate, guarantee, and uncover strategies for the rapid distribution of warning/signals to the public.

At subnational levels

1. District-level disaster management coordination and oversight is the responsibility of the District Disaster Management Committee (DDMC), which reports to the Deputy Commissioner (DC).

2. The Upazila Nirbahi Officer (UNO) presides over the Upazila Disaster Management Committee (UzDMC), whose job it is to oversee and organize disaster relief efforts in the Upazila.
3. The UDMC is tasked with coordinating, reviewing, and implementing the Union's disaster management efforts and reports directly to the Union Parishad's Chairman.
4. The Chairman of Pourashava (the municipality) presides over the Pourashava Disaster Management Committee (PDMC), which is responsible for coordinating, reviewing, and implementing disaster management actions within the municipality.
5. Mayor of City Corporations chairs the City Corporation Disaster Management Committee (CCDMC), which is responsible for coordinating, reviewing, and implementing disaster management actions within the City Corporation's authority.

Ministry of Disaster Management and Relief: Allocation of Business

The Ministry of Disaster Management and Relief's Disaster Management and Relief Division has received widespread acclaim for its innovative, timely, and sustainable approaches to disaster response. Natural disasters impair a nation's progress toward ending poverty and improving living standards. Through direct financial aid, food, and other forms of support, the Disaster Management and Relief Division helps to guarantee appropriate disaster management in the lead-up to and in the aftermath of disasters. Emergency response may lessen the impact of disasters, making it simpler to begin the recovery process. Key duties of the ministry are as follows:

1. General disaster preparedness and emergency response management, encompassing relief, rehabilitation, and safety net program development, assessment, and implementation.
2. Manage information systems for Vulnerable Group Feeding (VGF), Vulnerable Group Development (VGD), and other forms of social safety nets, including the maintenance of a database of program participants.
3. Programs for disaster relief, preparedness, risk assessment, and surveillance.
4. Personnel management, budgeting, and performance evaluation for this division's technical and non-cadre services, as well as those of its linked Directorates/Bureau and subordinate offices.
5. Management of disasters and relief efforts, including mitigation of potential hazards and organization of rapid responses, requires careful planning and execution of all involved tasks.
6. Implementing disaster risk reduction (DRR) across all relevant sectors, including central and local governments, non-governmental organizations (NGOs), community-based organizations (CBOs), and the wider civil society.
7. Putting into action plans and actions designed to mitigate the effects of climate change-related disasters.

8. Plans for disaster relief and rehabilitation, including the delivery of aid supplies, free assistance, and other forms of assistance, as well as their administration, coordination, and oversight.
9. Test relief, Vulnerable Group Feeding (VGF), Vulnerable Group Development (VGD), Food For Works Programme; Institutional Feeding Programme, Rural Infrastructure Maintenance Programme, Risk Reduction Programme, Road Maintenance Grants, House Building Grants, etc., are all examples of safety net programs that need approval, administration, and monitoring.
10. Multi-purpose disaster shelters, cyclone shelters, and flood shelters are built and maintained with the end goal of eliminating or greatly lowering catastrophe hazards.
11. Activities include the planning and implementation of different development projects/programmes on topics assigned to this Division.
12. Regional and international cooperation in the management of disasters, including the administration and coordination of aids, loans, grants, search-and-rescue operations, emergency relief, and technical help.
13. Tasks assigned to this section include liaising with international and regional entities, and negotiating relevant treaties and agreements with foreign nations and international organizations.
14. Evaluating the state of emergency and making a recommendation to declare a state of emergency, issuing evacuation orders, and keeping tabs on the early warning system are all part of disaster management.
15. Set up a national system to respond to disasters and continually enhance it.
16. The Standing Order on Disaster (SOD) is being revised and will be put into effect.
17. Create Emergency Operations Centres (EOCs) and other Disaster Management Information Centres (DMICs) at the national and local levels and ensure their smooth operation (EOCs).
18. Increasing the scope of earthquake and infrastructure collapse, Tsunami, fire, and other mass casualty event preventive and preparation measures.
19. Create, administer, and coordinate a program of disaster management training, development, and awareness-raising.
20. Financial, budgetary, and accounting concerns unique to this section.
21. Putting into action the many refugee-related initiatives.
22. All statutes and regulations pertaining to the purview of this department.
23. Questions and input on any of the responsibilities this Section has been given.
24. Costs associated with any area of jurisdiction assigned to this department, except judicial costs.
25. Everything else having to do with the Disaster Management and Relief Section of the Ministry of Food and Disaster Management.

4.4 How Does Coordination Work in Bangladesh? Evaluating the Effectiveness of Disaster Management Committees at Different Levels

Our research showed that proper coordination is an integral part of catastrophe preparedness and response. Disaster response coordination is challenging because of the needs present in both emergency and calm situations, including significant uncertainty, the need for quick decision-making and reaction, and time and resource limits. Coordination of emergency responses is difficult because of the many factors that must be taken into account in an emergency situation. These include the high degree of uncertainty that accompanies such situations, the rapidity with which they unfold, the possibility of widespread casualties, the lack of adequate preparation time, the magnitude of the resulting damage, the scarcity of available resources, and the disruption of essential infrastructure services like electricity, communication, and transportation.

Multiple authorities and large numbers of people being involved, conflicts of interest, and the pressing need for up-to-date information all add complexity to an already difficult situation. Any effective national effort to lessen the impact of disasters relies heavily on effective coordination between its many components. The core of DRR coordination is summarized in Table 4.1.

How do Standing Committees on Disaster Management Work?

Specific Roles and Responsibilities of Ministries & Government Offices as per SOD.

Prime Minister's Office

a. Ensure that catastrophe risk reduction is a top priority on national and local levels by issuing directions to the appropriate parties.
b. The Ministries and relevant government agencies and bodies approve/endorse the SOD as a legislative framework and the legal foundation for all of their operations and acts in accordance with their particular tasks and obligations as defined in the SOD.
c. Boosting efforts to establish and fortify national integrated catastrophe risk reduction measures.
d. Development policies, programming, and planning at all governmental levels should include risk reduction as appropriate, particularly in poverty reduction strategies and sectors and multi-sector policies and plans.
e. Adopting or revising, as needed, laws to assist disaster risk reduction, including rules and processes to encourage compliance and promote incentives for conducting risk reduction and mitigation measures.
f. funding the creation and implementation of disaster risk management policies, programs, rules, and regulations in all relevant sectors and authorities at all administrative and budgetary levels on the basis of clearly prioritized tasks.
g. Displaying the political will essential to promoting and integrating catastrophe risk reduction into development initiatives.

Table 4.1 Importance of coordination in disaster management

Why (challenges)	How (coordination support activities)
High uncertainty, sudden and unexpected events	• Real-time monitoring and timely alert notification for situation-awareness • Improvisation and rapid adaptation of predefined plans to the scenario • Periodical evaluation and update on existing coordination practices
Risk and possible mass casualty	• Threat, vulnerability, and risk assessment and countermeasure • Risk-sharing policy among parties involved • Operational sustainability management
Increased time pressure and urgency	• Repository of related plans, procedures, policies • Knowledge base and network of internal and external experts • Efficient information and intelligence mining; knowledge elicitation • Decision support technologies • Psychological fortitude to deal with effect and behavior
Severe resource shortage	• Policies (e.g., priority list and access control) for resource use and requisition • Logistic management and resource sharing network across local, national, and international levels • Self-equipped response teams • Law and order, price control mechanism management
Large-scale impact and damage	• Joint effort of governmental, public, and private sectors across local, national, and international boundaries • Broad information, intelligence, and resource sharing networks • An integrated public communication network to inform, guide, and reassure the general public
Disruption of infrastructure support	• Protection of critical infrastructures • Performance monitoring of built structures • Planning for infrastructure interdependencies such as proximity of foliage and civil infrastructure • Control of infrastructure redundancy • Management of alternative infrastructures
Multi-authority and massive people involvement	• Unified response command for coordination • Establishment of role structures with corresponding authority, responsibility, and accountability • Management of power and regulation conflict • Exercise of leadership and norms • Communication operability and interoperability

(continued)

Table 4.1 (continued)

Why (challenges)	How (coordination support activities)
Conflict of Interest	• Understand the political, ethnic, economic, and environmental impact • Shared vision and alignment of core interests • Reconcile the objectives of various involved parties in a politically sensitive and emotionally charged environment • Shared vision of priority among responders and the public
High demand for timely information	• Information gathering and provisioning • Information fusion and validation • Information exploitation and dissemination

Source Information gathered and complied by the researcher

h. Advocating for the adoption of the suggestions made by the NDMC, IMDMCC, EPAC, and NDMAC.
i. Instruct relevant Ministries to ensure that their personnel engage in disaster risk management activities by providing them with the necessary education, research, training, and awareness in the field of disaster management.
j. Manage the NDMC's operations as a whole.
k. Make sure that each year, the relevant Ministries evaluate whether or not their contingency plan is adequate and in line with the insights gained from the previous year's reaction, relief, and recovery efforts.
l. Activate all available resources in support of emergency, relief, and recovery activities by taking the initiative to issue instructions to the relevant Ministries.
m. Establish a framework to closely monitor continuing decisions and actions by the IMDMCC, EPAC, and NDMAC throughout response, relief, and recovery activities and ensure that the PM Office is represented at all such meetings.
n. The Disaster Monitoring and Coordination Cell will make it simpler to share and gather data in the event of a disaster.
o. Learn to communicate well with the DMIC.
p. Always be ready to lend a hand during emergency situations and assist with relief and restoration efforts if asked.
q. Make sure the accountable agency is sending out timely and accurate warnings.
r. Make that IMDMCC's efforts to coordinate between ministries during response, relief, and recovery have the resources they need to succeed.
s. Assist with relief, recovery, and rehabilitation operations and initiatives as needed by reallocating resources from the relevant Ministry, including personnel, to the impacted regions.

Cabinet Division

a. Plan for the adoption of the DM&RD, MoFDM, and IMDMCC-approved policy and legislative framework for disaster management.

b. Make it a requirement for all government agencies to include disaster preparedness measures into their long-term plans and initiatives.
c. Make it possible for yearly income and development budgets to include a block allocation for catastrophe risk reduction.
d. Will serve as the NDMC's secretary.

Armed Forces Division (AFD)

Following the discharge of regular duties, the division will adopt the following to put into motion its own action plans to deal with the calamities.

Risk Reduction

a. Prepare an earthquake contingency plan for the Armed Forces Division (AFD) and undertake a full sectoral risk assessment.
b. Set aside funds in the budget to carry out the strategy.
c. The most efficient use of funds requires a system of checks and balances to be set up.
d. Develop an earthquake preparedness policy and educate and train all employees on how to respond to earthquakes and other natural disasters.
e. Set up earthquake exercises and practice emergency response procedures.
f. Plan in advance how you will use search-and-rescue tools, infrastructure, relief supplies, and the safety of the impacted population.
g. Establish a plan to identify earthquake-prone sites and estimate potential losses.
h. Create a division-wide method for disseminating risk information.
i. Create a backup plan for its risk management and mitigation procedures per industry.

Emergency Response

Normal Times

a. Make a decision on who will act as the focal point for disaster management, and share that information with everyone involved.
b. Maintain constant communication and cooperation with the CCDR, NDRCG, and MoFDM.
c. Before the disaster season begins, make sure the three services are adequately equipped and have the necessary equipment to provide all the aid people in disaster-prone regions need for things like security, evacuation, and rescue.
d. Prepare a disaster management task force with the necessary skills and organize an emergency deployment force of the Armed Forces to carry out relief, rescue, and evacuation operations.
e. Create a strategy and get the required funding for catastrophe emergency operations to better react to the need and the unexpected.
f. Maintain funding for emergency planning and management.
g. Prepare plans for the military to assist civilian authorities based on requests and requisitions, working with the DMB, DM&RD.

h. Provide emergency reaction, relief, and recovery training for the Task Force Commander.
i. Keep all buildings, machinery, people, and materials secure.
j. Make a backup plan.

Alert and Warning Stage

a. Maintain constant vigilance and coordination through the Prime Minister's Control Room (24 h).
b. Ensure constant communication between the DM&RD, MoFDM's EOC, and the rest of the organization.
c. Maintain Army, Navy, and Air Force rescue, evacuation, and relief forces at the ready for deployment.
d. Confirm that plans are in place to deploy the Task Force to disaster-stricken regions as designated by the government.

Disaster Stage

(a) Deploy the military to assist in disaster response, relief, and rehabilitation efforts in response to a government request.
(b) Regularly update the Prime Minister's Office's coordination cell, as well as the IMDMCC and MoFDM, with data gathered on the rescue, relief, and rehabilitation efforts.
(c) Will expedite the Government's recovery, relief, and rehabilitation efforts.
(d) Provide the NDMC/DM&RD with a report detailing the military's efforts in the areas of relief and rehabilitation.

Bangladesh Army

During times of crisis, the Bangladeshi military plays a pivotal role in helping with rescue, aid, and recovery efforts. The Army Headquarters is responsible for the following during times of emergency.

Risk Reduction

(a) Conduct a thorough sector risk assessment of the Bangladesh Army in order to develop a strategy for mitigating and preparing for sector risks.
(b) Spending money on carrying out the strategy must be a priority.
(c) The most efficient use of funds requires a system of checks and balances to be set up.
(d) Carry out employee education and awareness training on disaster management problems.
(e) Set up a method of sharing information about potential dangers across industries.
(f) Create an earthquake-specific backup plan for its risk management and mitigation measures. Make sure the agency has a backup plan, and keep it up to date.

Emergency Response

Normal Times

(a) In both the Army's main headquarters and each division's main headquarters, designate someone to act as the primary point of contact for disaster management.
(b) Will make sure that smaller vehicles, trucks, rescue boats, and motorboats are set aside for times of crisis.
(c) Preparation for emergencies should consist of the following steps:
(d) Signals of caution and alertness.
(e) Methods of transmission.
(f) Testing and training for emergency situations.
(g) Securing buildings, machines, and people.
(h) In response to the civil administration.
(i) The locating of resources to support emergency response and relief efforts.
(j) Organization of one infantry company, one engineer company, and one medical company from the identified Task Force and reserve Task Force and Task Force (first aid, doctors, nurses, medicines).
(k) Develop and regularly update your own Action Plan to ensure the safety of people, property, infrastructure, and assets.
(l) Get ready for the call of civil administration in a crisis by creating an operational strategy.
(m) Conduct emergency response and recovery training for officers and jawans at all training institutions and staff colleges, including earthquake and fire simulations.
(n) Practice emergency procedures to make sure you're ready for anything.

Alert and Warning Stage

(a) The Director of Military Operations (Office and Residence), the control room and duty officers of the coordination cell of the Prime Minister's Secretariat, and the control rooms of the MoFDM, Navy, and Air Force should all have the phone number for the Disaster Control Room, which should be set up in the headquarters and concerned formation headquarters.
(b) If required, assign one Liaison Officer to the Prime Minister's Secretariat's coordination cell and keep them in touch with the MoFDM's operations center.
(c) Send out a warning to everyone involved. In locations where a crisis is likely to occur, you should issue instructions for the swift mobilization of the army and give them tasks.
(d) Form a Task Force in each formation to serve as a Crisis Management Team in case of an emergency. Each such group should have a complete infantry company, an engineer, sufficient first aid supplies, medical professionals with necessary supplies, and nursing aides.
(e) Reserve Task Forces should be activated if they become necessary.
(f) If necessary, reposition the Task Force to more advantageous locations.

(g) For purposes of evacuation, rescue, relief, healthcare, and rehabilitation, the Task Force Commander will work with local civic authorities.
(h) Regardless of what else may need to be done, prepare for the worse.
(i) Forward a detailed report on the situation and related actions to the Prime Minister's Secretariat's coordination cell.

Disaster Stage

(a) It is imperative that the Army and other relevant units have their disaster control room operating at all times.
(b) Gather data from disaster-prone areas on preparedness and other activities, and send it to the coordination cell of the Prime Minister's Secretariat and the EOC of the MoFDM for review by the National Disaster Management Coordinating Committee and the International Disaster Management Coordinating Committee.
(c) Coordinate Task Force deployment with the Prime Minister's Office and the Ministry of Defense, Land, and Mining.
(d) Help the government in the area by providing.
(e) Quickly leaving the building due to an emergency.
(f) SAR for a building that has just collapsed.
(g) Body and debris removal.
(h) Provide or arrange for access to medical care, including field hospital care, if needed.
(i) Avoiding the Spread of Disease.
(j) Overnight accommodation.
(k) Losses, repairs, and evaluations of requirements.
(l) The Prime Minister's Office coordination cell and the MoFDM EOC get frequent updates from Dispatch to keep the NDMC abreast of developments.
(m) Participate in anything else you see essential from a humanitarian perspective.
(n) Assist local authorities in executing rescue operations and conduct relief efforts in all impacted regions.

Early Recovery and Rehabilitation Stage

a. Assess the need for aid and reconstruction by conducting a survey in the impacted regions.
b. Collaborate with authorities in the region to clean up the mess and restore the area's natural beauty.
c. Help organizations focused on preventing epidemics while also assisting those providing medical care to those who have been hurt in the disaster zone.
d. Give the local government a hand in securing clean water for consumption.
e. If necessary, establish mobile medical clinics.
f. Provide aid to government in building emergency housing.
g. If the government asks for your help, you must take part in any restoration or rebuilding efforts.
h. Do all you can, from a humanitarian perspective, to help the people who have been impacted.

i. Send the comprehensive report on the status of relief and rehabilitation efforts to the coordination cell of the PM's office and the EOC of the MoFDM for review by the NDMC. Assess the need for aid and reconstruction by conducting a survey in the impacted regions.
j. Collaborate with authorities in the region to clean up the mess and restore the area's natural beauty.
k. Help organizations focused on preventing epidemics while also assisting those providing medical care to those who have been hurt in the disaster zone.
l. Give the local government a hand in securing clean water for consumption.
m. If necessary, establish mobile medical clinics.
n. Provide aid to government in building emergency housing.
o. If the government asks for your help, you must take part in any restoration or rebuilding efforts.
p. Do all you can, from a humanitarian perspective, to help the people who have been impacted.
q. Send the comprehensive report on the status of relief and rehabilitation efforts to the coordination cell of the PM's office and the EOC of the MoFDM for review by the NDMC.

Bangladesh Navy

When cyclones strike, the Bangladesh Navy will do all in its power to help the coastal governments it serves. The breadth of their typical resources and boats should be sufficient for this kind of aid. As part of this aid, conveyance of emergency supplies from Chittagong to offshore islands and from Narayanganj/Khulna to Barisal/Patuakhali is required (only in those areas where depth of water is sufficient for such transportation). These vessels will carry medical personnel, both military and civilian, and supplies supplied by or arranged for by Deputy Commissioners. Maintain readiness to carry out any emergency government task in the case of a catastrophic cyclonic catastrophe. The Bangladesh Navy will carry out the following tasks in addition to carrying out their own work plans.

Risk Reduction

(a) Conducting a thorough sector risk assessment of the Bangladesh Navy would help in developing a strategy to mitigate risks and be prepared for them.
(b) Spending money on carrying out the strategy must be a priority.
(c) The most efficient use of funds requires a system of checks and balances to be set up.
(d) Educate and raise awareness among your workforce. Make sure that everyone working there has had earthquake training.
(e) Departmental exercises should be held annually to assess preparation and the status of the department's contingency plan (particular attention might be paid to cyclones).
(f) Take an inventory of the office space and provide the results to the DMB, DM&RD.

(g) Set up a method of sharing information about potential dangers across industries.

Emergency Response

Normal Times

(a) At Naval Headquarters, Khulna, and Chittagong, the Navy will establish three Disaster Management Focal Points.
(b) Prepare the Naval Force for responding to and recovering from disasters.
(c) Preparation for emergencies should consist of the following steps:
(d) Signals of caution and alertness.
(e) Mechanics of conveying messages.
(f) Contingency planning drills and evaluations.
(g) Protection for buildings, vessels, machines, and people.
(h) In response to the government.
(i) Locating resources to support emergency response and relief efforts.
(j) Finding the most efficient water transportation options.
(k) A regional command center in Dhaka, a regional command center in Khulna, and a regional command center in Chittagong have been established.
(l) Make the necessary institutional preparations for emergency long-term relief operations.
(m) In order to receive warning signals and make the necessary preparations in advance, the Operations Directorate of Naval Headquarters will maintain communication with the Meteorological Department/Ministry of Food and Disaster Management.
(n) You should compile a list of available watercraft for use in relief and rescue efforts before a storm hits.
(o) Rescue and rehabilitation training for employees.
(p) Use yearly exercises to evaluate and refine your organization's disaster management response strategy in light of the SOD.
(q) Make contingency measures to safeguard the Bangladesh Navy's vessels, machinery, infrastructure, supplies, and people.

Alert and Warning Stage

(a) The command to sound the alarm should be sent to all troops.
(b) Chittagong's Sea Level Data Monitoring unit will gather data, share it regularly with the IOTWS and BMD, and respond appropriately to any emergencies that may arise.
(c) The Prime Minister's Coordination Cell, the Emergency Operations Center at the Ministry of Defense and Disaster Management, and the Control Rooms of the Bangladesh Army and the Bangladesh Air Force should all have the contact information for the Navy's control room, control room, and duty officer in case of a disaster.
(d) You should make preparations to safeguard the vessels, infrastructure, people, and equipment of the Bangladesh Navy that may be at risk from the calamity.

(e) One Liaison Officer should be appointed to the Prime Minister's Coordination Cell in order to maintain communication with the EOC at the Ministry of Defense and the Control Room of the Bangladesh Army and Air Force.

Disaster Stage

(a) The government has asked that you send ships to help in their duties.
(b) Help the government in the area by providing:

    i. Losses, repairs, and evaluations of requirements
    ii. Health care
    iii. Moving and distributing aid supplies
    iv. Resumption of Off-Shore and Communications Operations

(c) Set up enough staffing in the command centers at the Naval Headquarters in Chittagong and Khulna.
(d) Await further orders to begin cyclone relief activities, and keep all ships and stations at the ready.
(e) The Navy's HQ must maintain constant communication with the EOC and Control Room at the Ministry of Defense and Security.
(f) In close consultation with the relevant government agency, implement all measures of security that may reasonably be expected to achieve their intended purpose.
(g) The ships and stations will help the local government when ordered to do so by Naval Headquarters.
(h) It will report its progress to the MoFDM's EOC and the coordination cell of the Prime Minister's office on a regular basis.

Early Recovery and Rehabilitation Stage

(a) Conduct a damage, loss, and needs assessment (DLNA) in the impacted regions and provide advice to AFD and DMB on how to proceed with relief and rehabilitation efforts.
(b) Help the local government in any manner you can with relief and restoration efforts in the impacted regions.
(c) Help the government hand out supplies to those who need them.
(d) Help the local government in whatever way you can until things go back to normal.
(e) Compile and submit to the NDMC comprehensive reports covering all areas of the relief and rehabilitation efforts.

4.4.1 Bangladesh Air Force

Bangladesh Air Force plays a crucial part in national catastrophe response. The Bangladesh Air Force must be ready to respond to any emergency, including those caused by natural disasters. The following tasks, in addition to its regular ones, will be carried out by it.

Risk Reduction

(a) Conduct a thorough evaluation of the risks facing the Bangladesh Air Force in order to develop a strategy for mitigating those risks and being ready for them.
(b) Spending money on carrying out the strategy must be a priority.
(c) The most efficient use of funds requires a system of checks and balances to be set up.
(d) Get the Ministry to set up a framework for communicating risks across different industries.
(e) Make sure the agency has a backup plan, and keep it up to date.
(f) Educate and raise awareness among all employees by giving them training on how to handle earthquakes and other disasters.
(g) Operationalize the contingency plans via staging yearly exercises on preparation and amend the plan as necessary.

Emergency Response

Normal Times

(a) Air Force bases and headquarters should each appoint a single Disaster Management Focal Point and share that person's information with the DMB and any other relevant parties.
(b) Updated aerial maps of flood-prone areas, coastal regions, and offshore islands should be kept in the Disaster Control Room. Don't let any information regarding airports and helipads become out of current.
(c) Establish reliable protocols in advance for receiving weather alerts from the Meteorological Department and other sources.
(d) Create emergency plans to safeguard aircraft, equipment, and facilities, and to react to requests from civil authorities, and draft orders to implement these preparations.
(e) Learn how to respond to disasters and aid in their recovery by enrolling in a relevant training program.
(f) Preparation for emergencies should consist of the following steps:

   i. Signals of caution and alertness
   ii. Methods of transmission
   iii. Testing and training for emergency situations
   iv. Installation, airplane, tool, and manpower safety
   v. in response to the civil administration
   vi. The locating of resources to support emergency response and relief efforts
   vii. Airfields and helipads that will work may be located
   viii. Aircraft id for transport and reconnaissance.

Alert and Warning Stage

(a) Send out safety preventative instructions to everyone involved.
(b) Maintain a ready supply of transport planes and helicopters.
(c) Bangladesh Air Force aircraft and equipment in cyclone-prone locations should be moved to a safer location as a matter of urgency.
(d) Maintain communication with the Emergency Operations Center at the Ministry of Defense and the Army and Navy Control Room by assigning one Liaison Officer to the Prime Minister's Coordination cell as needed.
(e) Give the Emergency Operations Center at the Ministry of Foreign Affairs and Defense, the Control Room of the Prime Minister's Coordination Cell, and the Control Room of the Army and Navy the contact information for the Director of Air Operations (Office and Residence) of control room and the duty officer.

Disaster Stage

(a) The cyclone/flood scenario requires continual monitoring.
(b) Supply the relevant authorities with any supplementary meteorological data, if gathered independently of BMD and BWDB.
(c) Rescue efforts and damage assessments after cyclones need the readiness of transport aircraft and helicopters for initial trial flights as soon as weather conditions permit.
(d) In the event of extreme flooding, aid relief efforts by airlifting people as needed by civic authorities.
(e) Allow the military to utilize the airport for temporary deportation of relief supplies and emergency search-and-rescue operations.

Rehabilitation Stage

(a) Send a report detailing your aerial survey's results to the National Disaster Management Coordination Center (NDMC) and the International Disaster Management Coordination Center (IMDMCC).
(b) Severely wounded people should be sent to nearby hospitals for treatment.
(c) Donate transport planes and helicopters so that authorities may fly over the impacted regions and assess the situation.
(d) Drop off supplies like food and water to the impacted communities by airdrop if necessary.
(e) Help the government provide relief supplies, medication, and medical personnel to the impacted region by chartering helicopters.
(f) Use the BAF's wireless, radio, night vision equipment, and telephone communications as a supplement to the regular civil communication system for the transfer of vital information and data.
(g) Take on whatever additional responsibilities the government may assign in the sake of the broader relief effort.
(h) Bring together a comprehensive report covering all that has been done so far to aid those in need.
(i) Please assign one Liaison Officer to coordinate with foreign air forces on relief missions.

4.4.2 NGO Affairs Bureau

Risk Reduction

(a) The FD-6 Format should be updated to take into account disaster risk reduction measures, and thus guidelines should be issued to that effect.
(b) It is imperative that NGOs be instructed to include disaster management messaging into their various skill and awareness campaigns.
(c) You should tell the NGOs that operate in the city to make a strategy for dealing with an earthquake and to practice good management.
(d) Assuring DMB's participation in NGOs' DRR initiatives.

Emergency Response

Normal Times

(a) One member of the senior staff should be designated as the Disaster Management Focal Point and should attend the IMDMCC and NGO coordination committee meetings hosted by the MoFDM.
(b) Create and update a database of non-governmental organizations (NGOs) with humanitarian initiatives.
(c) Connect the NGOs with open lines of communication.

Warning/Alerting Stage

a. Instruct NGOs to disseminate the warning signal to the communities.

Disaster Stage

(a) Request that the DDMC and the UzDMC get the necessary cooperation and assistance from all non-governmental organizations.
(b) Tell non-governmental organizations to help the government at the local level by providing:

   i. Injured individuals are being evacuated to safer areas
   ii. Analyses of damages and need
   iii. Health care
   iv. Moving and distributing aid supplies
   v. Communications
   vi. Microcredit installment collections in the impacted regions
   vii. Help expedite the release of foreign humanitarian commodities held by NGOs at an airport or seaport.

Rehabilitation Stage

(a) Create and implement a standard operating procedure to coordinate early recovery and rehabilitation efforts among the government, international, national, and local non-governmental organizations, and other stakeholders.

(b) NGOs' early recovery and rehabilitation efforts should be coordinated.
(c) Keep track of the total amount contributed by various actors in different regions.

Ministry of Food and Disaster Management (MoFDM)

The Government's main point of contact for all matters pertaining to food and disasters is the Ministry of Food and Disaster Management (MoFDM). Concerns about disaster management will be handled by the DM&RD.

Disaster Management and Relief Division (DM&RD)

DM&RD is in charge of dealing with the disaster management concerns. During the predisaster, disaster, and post-disaster recovery phases, the DMB will supply the DM&RD and Ministry with all relevant data. The NDMC and IMDMCC may rely on the division to offer them with data and guidance as they work through issues. The Secretary of the DM&RD, MoFDM will be in charge of overseeing the actions of all government officials participating in disaster relief efforts.

Risk Reduction

(a) Appoint a seasoned employee to serve as the DM&RD's primary contact.
(b) In order to achieve the following strategic objectives in the aftermath of a catastrophe in Bangladesh, it is necessary to develop and implement national policy, planning, and legislative frameworks.

    i. Bringing the Disaster Management System up to Professional Standards
    ii. Combining efforts to lower risks
    iii. Improving Institutional Structures
    iv. Supporting vulnerable populations
    v. Increasing sector-wide, hazard-specific, and overall risk reduction initiatives
    vi. Improving the reliability of emergency services
    vii. Developing and strengthening networks

(c) It is important to create a new SOD and evaluate and update the National Plan for Disaster Management on a regular basis.
(d) Create detailed instructions and sample documents for incorporating disaster management ideas and practices into national development planning procedures across risks, sectors, and administrative levels.
(e) Assessments of hazards, risks, and vulnerabilities must be carried out on a national, district, Upazila, and union level, and risk reduction action plans (RRAP) and strategies for putting them into effect must be developed.
(f) The Upazilas, unique disaster-prone locations inside those Upazilas, and the people living there should be mapped out in order to determine the extent of the potential damage.
(g) Will evaluate the interministerial earthquake contingency strategy every six months.
(h) The EOC will be ready to respond to any earthquake-related crises.

(i) Will make sure that all parts and components of earthquake risk maps are readily available.
(j) Make sure the Building Codes are being followed correctly by coordinating with the city's planning department and the Ministry of Public Works.
(k) Action plans for reducing earthquake risk and emergency backup plans should be made.
(l) Bring the government's efforts to mitigate risk together across departments.

Emergency Response

Subject to the approval of the Minister, the Secretary, DM&RD will take decisions on the following matters.

(a) Relax any Standing Regulation (Standing Order) on relief distribution in the public interest.

Normal Times

(a) At least once every 3 (three) months, the Ministry of Disaster Management should review its own Action Plan.
(b) Determine the Upazilas that will be affected by the catastrophe, as well as the specific disaster zones within those Upazilas and the populations who will likely be impacted by the disaster.
(c) Keep up-to-date a list of international and non-governmental organizations that are ready to help with disaster relief, recovery, and prevention efforts.
(d) Keep accessible data on food, relief supplies, and transportation for use in the event of a crisis.
(e) Direct everyone responsible for ensuring availability of the SOD at village, Union, Upazila, and District DMCs and to stakeholders.
(f) Schedule NDMC and IMDMCC meetings to evaluate the disaster readiness of various government agencies, municipalities, autonomous organizations, CPPs, BDRCSs, NGOs, etc.
(g) Make sure the Ministry has a constant line of communication with the Upazila and District Offices.
(h) Issue required instructions for efficient coordination of measures linked to catastrophe and response.

Alert and Warning Stage

(a) Command the transportation of aid supplies to where they are needed.
(b) Choose a Ministry Point of Contact and make sure everyone knows their name and number.
(c) DMB should be instructed to activate the EOC at MoFDM and establish its own EOC, as well as direct the opening of a Control Room for all levels of disaster-related activity.
(d) Maintain open lines of communication with BMD and BWDB and issue orders to collect inputs (alert message and signal) from the unit.

(e) Make sure that all available channels of mass communication (including the airwaves, the fax machine, the phone, the Internet, and mobile phones) are used to spread alerts. Notify the CPP, the BDRCS, the NGOs, the Deputy Commissioners, and any other relevant authorities or officials.
(f) Keep the EOC open 24 h a day.
(g) Get the CPP Implementation Board together and let everyone know what they decide.
(h) Notify the Prime Minister, the Chairman of the IMDMCC, and the Chairman of the NDMC of the catastrophe circumstances and the measures taken to address them.
(i) Do your best to get the District, Upazila, and Union DMCs together whenever necessary.
(j) I am writing to ask that the AFD maintain a helicopter and transport plane for the purposes of loss and damage assessment and rescue missions.
(k) Ask the AFD and the BIWTC to maintain watercraft for use in search-and-rescue and humanitarian efforts.
(l) Command people to leave for safer areas so that lives and property may be preserved.
(m) Call on the Army through AFD to be immediately deployable to the affected areas.
(n) The Chairs of DMCs (Divisional Commissioners, Deputy Commissioners, Upazila Nirbahi Officers, and Union Parishad Chairmen) and other interested authorities should be made aware of the very alarming warning signs and the necessary preparations to be done.
(o) Give the order to the District and Upazila administrations to call for the necessary transportation for the rescue and relief efforts.
(p) Set up times and places for the NDMC and IMDMCC to meet.
(q) Instruct local government to assess potential victims and relocate them to safer areas in conjunction with the BMD.
(r) Make sure that the Bangladesh Betar (Radio) and the BTV are used to continuously send out alerts.
(s) Maintain constant contact with your District Control Room, CPP, and BMD.
(t) Get the disaster-stricken communities the supplies they need in advance.
(u) Choose one person to serve as the Liaison Officer between the Prime Minister's office and the coordination cell.

Disaster Stage

(a) I am writing to ask that the Bangladesh Navy and Bangladesh Air Force have their ships and aircraft available for damage assessment and rescue operations as soon as the weather permits.
(b) Make an appeal to the government for help with relief and rescue efforts by the AFD.
(c) Help rescue efforts go smoothly by coordinating with non-governmental organizations.
(d) Set up gatherings of the NDMC and the IMDMCC.

(e) Take notes on any damage or losses.
(f) Determine what other resources will be needed to help with relief and rebuilding efforts.
(g) Assemble finances and supplies for free aid as soon as possible.

Rehabilitation Stage

(a) Set up programs like the Housing Construction Grant, Test Relief, and Food for Work.
(b) Maintain emergency rehabilitation efforts in impacted regions until life can return to normal.
(c) Manage a network of rehabilitation centers.

Food Division

The Food Division plays a crucial part in ensuring that disaster-prone communities always have enough food on hand in the event of an attack from a natural catastrophe.

Risk Reduction

(a) Adopt national food security strategies, plans, and programs that take catastrophe risk reduction into account.
(b) When estimating food supply needs, it's important to factor in both potential and actual disasters.
(c) Assess the central, district, and outlying offices' susceptibility to disaster by conducting vulnerability and risk assessments of food infrastructure facilities and then taking appropriate action, such as retrofitting, as well as creating a contingency plan. When planning for new storage facilities and infrastructures, keep in mind potential threats and BNBC.
(d) Improve security at food storage facilities and keep them well-maintained to reduce the likelihood of theft or spoilage.
(e) Send the appropriate instructions and directives to the relevant divisions to guarantee the food division's risk reduction strategies are put into action.
(f) Maintain financial backing for the government's agenda of social safety net.

Emergency Response

Normal Times

(a) Find someone to be the Control Room's point person for disaster management.
(b) Choose one officer to serve as the Liaison and maintain contact with the DM&RD's Control Center.
(c) It is recommended that DG Food keep databases of vehicles, water transports, country boats, etc., and the owners' identities and the amount of commodities held in each godown updated. Examine how well each subordinate department is doing with its quarterly work plan.
(d) Protect food and other items kept in food godowns by moving them to more secure locations.

(e) Make that the distribution of VGF, VGD, and other test relief supports is well-coordinated with DRR.

Alert and Warning Stage

(a) Get the Division's Control Room up and running.
(b) One Liaison Officer should be appointed to serve as the point of contact between the DM&RD's Emergency Operations Center and the rest of the organization.
(c) Maintain a reasonable price for grains used in cooking.

Disaster Stage

(a) Maintain a constant connection with the division's EOC and run your own control room around the clock.
(b) Daily reports of impacted regions should be gathered and sent to the IMDMCC from reliable sources.
(c) Make sure there is enough food for the people living in the impacted regions by running a specific rationing system and open market sales if necessary.
(d) Avoid price manipulation and stockpiling by taking preventative measures to keep commodity costs low.

Rehabilitation Stage

(a) Quickly analyze the extent of the damage to the Food Division's infrastructure and services, and then take any necessary corrective action.
(b) Arrange for suitable plans to be prepared and funding to be allocated for the purpose of repairing and rebuilding storage facilities.

Disaster Management Bureau (DMB)

Under the Ministry of Disaster Management, the Disaster Management Bureau was established in 1992. This section details the roles it plays. The DMB will keep track of a variety of cluster groups and their duties, as well as the current response status, including any budget shortfalls, intercluster coordination efforts, or other relevant information.

Risk Reduction

(a) Help the government with any questions they may have about catastrophe planning.
(b) Legislation on disaster prevention and relief should be proposed, together with any other relevant directives.
(c) Keep in contact with and coordinate the efforts of various government agencies, humanitarian organizations, non-governmental organizations, and voluntary organizations involved in disaster management.
(d) In order to lessen the impact of natural disasters like cyclones, floods, and earthquakes, you should draft some rules or instructions.
(e) Working with the Planning Commission and other relevant authorities, draft guidelines for disaster mitigation and mainstreaming disaster risk reduction.

(f) Help the government, local government, and non-governmental organizations prepare for and respond to disasters like cyclones, floods, and earthquakes by sending your officials and elected representatives to a training session.
(g) Help the Geological Survey of Bangladesh out by sharing maps showing where earthquakes are most likely to occur.
(h) Create atlas of cities most at danger from earthquakes.
(i) Make sure the government is aware of the need for search-and-rescue equipment and transportation/vessel in the event of an earthquake, storm, flood, or tsunami.
(j) Acquire and disperse search-and-rescue gear to the appropriate authorities for use in the aftermath of natural disasters including earthquakes, cyclones, floods, and tsunamis.
(k) After the first "procurement of equipment for search-and-rescue operation for earthquake and other catastrophes" phase is complete, we will go forward with the necessary measures to extend the project as mandated by the government.
(l) Using the Emergency Cyclone Recovery and Restoration Project from 2007 and other suitable initiatives and activities, make sure that DRR is being practiced throughout the coastal areas of the nation and that persons impacted by the cyclone have access to the means to support themselves.
(m) Prepare for earthquakes by doing yearly drills and coordinating with appropriate first responding organizations/DMCs to conduct simulated exercises on cyclones, floods, and other disasters, and then reviewing the readiness status/position.
(n) Study issues including earthquake catastrophe financing, drought prevention, ways to make ends meet, and more.
(o) Investigate how people are adapting to new climate-related risks.
(p) Aid, coordinate, and keep an eye on the necessary agencies to make sure the Building Codes are being followed as intended.
(q) Guidelines containing ideas and recommendations for house-building projects should be formulated after analyzing the risk maps to be provided by the Geological Survey of Bangladesh.
(r) Develop a strategy for minimizing exposure to earthquake hazards and a set of guidelines for making preparations in the event of an earthquake.
(s) Carry all measures necessary to lessen the impact of a tsunami.
(t) Contribute to the Ministry's Efforts in:

    i. Development of Bangladesh's first national policy, planning, and legal frameworks for disaster management
    ii. The District, Upazila, and Union levels will develop a framework for a disaster management action plan and put it into practice.
    iii. Planning for disasters at all levels of government requires the creation of precise standards and templates that can be used to help integrate disaster management ideas and practices.
    iv. Improving government's ability to reduce risk across sectors.

Emergency Response

Normal Times

(a) Take on a variety of initiatives aimed at informing the general public, government workers, and professionals on how they may help reduce catastrophe risks.
(b) Assist the NDMAC with disaster-related secretarial tasks.
(c) Facilitate the dissemination of the SOD, the National Disaster Management Plan, and any other relevant recommendations.
(d) Train government workers, elected officials, and others in crisis management in conjunction with several Ministries, local authorities, training institutes, academies, and non-governmental organizations.
(e) Construct a national Emergency Operations Center (EOC) and disaster management information center (DMIC) with enhanced communication capabilities and use it to share information with public and private organizations.
(f) Give out reading materials, maps, and other resources to those working in disaster management.
(g) Keep an eye out for roadblocks that might prevent the Action Plan or project from being completed, and keep the DM&RD updated on the potential impacts on people and property, as well as on disaster preparation, response, and mitigation.
(h) Gather and save data on any structures built above flood level, such as cyclone shelters, embankments, and platforms (flood-proofing).
(i) Relate the efforts of NGOs in emergency management.
(j) Make sure there are campaigns to educate the public about weather warnings.
(k) Set up periodic radio and television public service announcements to educate the public on different disaster-related problems.
(l) Through the help of concerned DMCs and other groups, plan meetings, seminars, and workshops at the national, District, Upazila, and Union levels to raise awareness of the issue of natural catastrophes.
(m) Liaise with the Department of Education and the Department of Primary and Mass Education to include disaster education into K-12, tertiary, and tertiary-adult education programs.
(n) Help get the word out about cyclone alerts in the local community using posters, cultural events, documentaries, and more!

Alert and Warning stage

(a) Ensure that relevant authorities, agencies, and the media are receiving early warning signs of impending catastrophes.
(b) Help the DM&RD with emergency programs in high-risk locations involving several organizations and individuals who live there.
(c) Get the word out and start a national awareness campaign about the new early warning system.
(d) As soon as Signal No. 4 is sent, make sure Gano Durjog Barta gets all the attention it deserves.

(e) Install a local tsunami and storm surge warning (siren) system.
(f) Initiate Control Room operations and maintain communication with other agencies to ensure the success of their Action Plan.
(g) Send out regular briefings to foreign embassies and UN Missions throughout the tragedy.
(h) Tell the local government to figure out how much aid is needed once the harm has been done.

Disaster Stage

(a) A constant EOC presence is required (24 h).
(b) Help the DM&RD organize teams to conduct an initial evaluation of damage and requirements.
(c) DMB should set up a permanent cell to analyze damage, loss, and needs.
(d) Help the IMDMCC coordinate efforts between the government, NGOs, and other groups to provide aid and rebuild the affected areas.
(e) Keep tabs on how things are doing with the rescue, relief, and rehabilitation efforts, pinpoint any issues or needs, and bring them to the attention of the right people.
(f) Help the Economic Relations Division, the Ministry of Information, international agencies, non-governmental organizations (NGOs), and other relevant parties get the data they need from the DM&RD.
(g) Daily news bulletins should be sent to foreign embassies and UN offices, as well as all international organizations including the World Bank.
(h) Report back to the Ministry on how the supplies are being put to use in the rescue and recovery efforts.
(i) See to it that the SAR apparatus you've acquired and distributed is put to use.

Rehabilitation Stage

(a) Provide data/input to the relevant authorities so that a rehabilitation strategy may be developed.
(b) Incorporate measures to reduce the likelihood of more disasters occurring as part of the restoration process.
(c) Prepare reports based on the experience/knowledge gained in this regard, publish the same, and make necessary changes in the training programmes and future policy based on the findings of the post-mortem of the steps taken in the overall management in the emergency response to the disaster.

Directorate of Relief and Rehabilitation (DRR)

The following are further responsibilities that will fall within this Directorate's remit.

Risk Reduction

(a) Create standards and practices for evaluating community exposure to and risk from hazards.

(b) It is important to conduct hazard, risk, and vulnerability assessments and mapping at the national, District, Upazila, and Union levels to pinpoint the Upazilas and particular disaster zones under such Upazilas, as well as the population that would likely be impacted by a catastrophe.
(c) Create programs for the vulnerable, such as a safety net, that will aid in reducing risks in the community, and make sure your organization has a plan in place in case something goes wrong.
(d) Maintain a schedule of frequent training sessions for the chosen volunteers to help them become ready for earthquakes.
(e) Arrange for the DRR personnel who will be handling earthquakes to attend training on how to prepare for and handle such an event.

Emergency Response

Normal Times

(a) Make sure disaster-prone places have appropriate supplies and that they are stocked, secured, and well-maintained.
(b) Build roads to safe areas, plant trees, and build shelters using the resources you get via the Food for Works Program.

Alert and Warning Stage

(a) Establish a Command Center inside the Agency and keep in contact with the Ministry's Emergency Operations Center.
(b) Warn all authorities to stay on guard.
(c) The Ministry should be made aware of the status of relief efforts in impacted regions.
(d) Keep the Ministry updated on the situation every day.
(e) Maintain records in the LSD and CSD of the affected regions detailing the amount of relief supplies and food grains currently at hand.

Disaster Stage

(a) Bring aid to the afflicted regions as soon as possible.
(b) Send out orders to the field to assist the local government with any rescue or evacuation efforts.
(c) Put the watercraft at the disposal of the proper authorities so that they may be used to move aid supplies and conduct evacuation and rescue operations for the affected population.
(d) Keep the Ministry informed of the needs for emergency supplies.
(e) Promptly notify the Ministry of any need for exceptional assistance.
(f) Maintain accurate records of how aid money is used.

Rehabilitation Stage

(a) Make a recommendation to the Ministry for the distribution of relief supplies after visiting the affected areas and determining the extent of the disaster.

(b) Quickly distribute funds for home construction, free aid, and other resources as specified in the delegation of authority.
(c) Provide input to the Ministry about the distribution of (required) housing construction grants, test relief, gratuitous relief, and other materials outside the scope of delegated authority.
(d) Put out the necessary government decree.
(e) Keep up the most important restoration efforts.
(f) Bring the government up to speed on all of your expenditures by submitting a consolidated report.

Duties of DRR Officials at District and Upazila Level

Under the direction of the chair of the concerned DMC, the DC, and the UNO of the relevant Upazila, the DRRO and PIO will be responsible for the following tasks within their districts (UNO).

Risk Reduction

(a) Form Upazila and Union Disaster Management Committees and provide training in all aspects of disaster management, with a focus on earthquake preparation.
(b) Create a standardized method of assessing dangers and weaknesses at the neighborhood and family levels.
(c) The Union, Pourashava, Upazila, and District Disaster Management Committees (DMCs) need help from the government and NGOs to put together and carry out their plans to mitigate disasters.
(d) Make sure the government and nonprofits are working together to mitigate risks.
(e) Maintaining communication with the DMB will allow you to schedule frequent seminars and trainings on disaster-related topics.
(f) Ensure that the disaster risk elements and the potential for risk reduction have been thoroughly evaluated in the planning and implementation of the District and Upazila development programs.
(g) Assist District, Upazila, and Union Disaster Management Committees in establishing locally appropriate funding for enacting risk reduction action plans.
(h) As soon as a disaster forecast or warning is issued (for things like a tornado, cyclone, tidal surge, earthquake, landslide, river erosion, tsunami, heavy rainfall, no rainfall, drought, flood, water logging, high tide, cold wave, etc.), it must be widely disseminated among District officials and relevant individuals/organizations, and then measures must be taken to get the messages to the appropriate people in the Union, Pourashava, and Upazila.

Emergency Response

Normal Times

(a) Keep aid supplies safe.
(b) Evacuation supplies must be checked for fitness.

(c) Make localized backup plans just in case.

Alert and Warning Stage

(a) Make plans for the launch of the command center at the UNO and DC headquarters.
(b) Maintain communication between the catastrophe center and the UNO, including updates on the status of disaster-prone regions.
(c) Obtain the DC's permission before submitting a request to the Directorate of Relief and Rehabilitation to increase the supply of relief goods and relocate the contents of warehouses at risk of being damaged.
(d) Assist District and Upazila authorities in coordinating the efforts of various non-governmental organizations.
(e) Don't let your guard down when it comes to the safety of aid workers on watercraft.
(f) Examine the located shelters and submit your findings to DMB and DRR.

Disaster Stage

(a) Complete plans for transporting aid supplies to recipients.
(b) Please lend a hand with the rescue efforts.
(c) Determine the extent of the casualties and damage, and transmit your findings to the UN and Washington, DC.
(d) Assist the Union authorities in supervising and monitoring the delivery of relief supplies as approved.

Rehabilitation Stage

(a) Please notify the DC, DRR, and DMB of any losses, damages, and requirements.
(b) Move quickly to provide those impacted by the disaster their grants to construct homes, any gifts they may be eligible for, and any other relief supplies they may need.
(c) Send report on relief and rehabilitation efforts to DRR.
(d) Do your best to always have the assessment reports of relief efforts prepared for inspection.
(e) Arrange employment of disaster-affected persons via test relief.

Cyclone Preparedness Programme (CPP)
Cyclone Preparedness Programme (Headquarters, Dhaka)

Risk Reduction

(a) Make sure you have a representative at the National Disaster Management Council (NDMC) and the International Disaster Management Coordinating Committee (IMDMCC).
(b) CPP policies, plans, and programs should take catastrophe risk reduction into account.
(c) Make use of national disaster-prevention committees in your work.

(d) Provide funding for risk management initiatives and provide safety training for employees and volunteers.

Emergency Response

Normal Times

(a) Programs for disaster preparation should be organized on an ongoing basis, and exercises should be held annually in April and September to evaluate progress.
(b) Make it a yearly tradition to start seeking out volunteers and getting them trained before April.
(c) Establish committees at the Unit, Union, and Upazila levels.
(d) Make that Union and Upazila Offices are set up, and that meetings are held at the Union and Upazila levels.
(e) Make sure that volunteer group leaders get and utilize warning signal kits.
(f) Make sure there is constant wireless contact between the CPP's main office and the Upazila office, and the Upazila office and the Union office.
(g) Provide guidance to DMCs at the local level on how to choose shelters and high-rise safe locations, and disseminate this information to the public.
(h) Increase public understanding of cyclone warnings and DMB's role in emergency planning through town hall meetings, educational materials, and media productions.
(i) Maintaining constant fax contact with the weather service requires constant attention to detail.

Alert Stage

(a) Help the District, Upazila, and Union governments set up control rooms by establishing them at CPP headquarters and in regional offices.
(b) Keep in constant contact with BMD and the DMIC.
(c) As soon as special weather bulletins are issued by BMD, they should be sent to the Upazila and regional offices, and from there the Union offices should be instructed to receive them.
(d) Tell the participants to tune in to the radio to hear the regular news or the CPP broadcasts. To communicate with other union group leaders through radio or liaison volunteers, and to recommend to union group leaders and their coworkers that they order the volunteers to begin work in accordance with the criteria outlined in the cyclone instructions booklet released by CPP.
(e) Informing the Chairman and the Members of the CPP Implementation Board on the cyclone's progress is essential.
(f) Do not hesitate to contact the District Collectors, Upazila Executive Officers, Union Parishad Chairman and members, and local non-governmental organizations.

Warning Stage

(a) It is suggested that the CPP Implementation Board have an immediate meeting.
(b) Find out whether the CPP volunteers have begun working on the ground.

(c) Advise individuals on how to safely relocate their livestock, poultry, and other domestic animals to elevated land and killas.
(d) Advise Chairman of UzDMC to convene meeting by notifying CPP Development Officer.
(e) Commission the Development Officer to convene the Union DMC.
(f) Take action on the choices made during the Board of Implementation's impromptu meeting.
(g) To ensure that the choices made by the Upazila and Union DMCs are carried out, it is imperative that you instruct the Development Officer to work with you on this.
(h) Get those particular weather reports to the regional, Upazila, and union offices through Wi-Fi.
(i) Do your best to alert the public (about the calamity) through the Development Officer and the volunteers.
(j) Be sure to keep the Chairman of the Implementation Board, the Director General of the Disaster Management Board, and the Directors of Disaster Risk Reduction and the Business Continuity and Continuity Services updated on the situation at all times.

Disaster Stage

(a) Align its efforts with those of other relevant government and non-government organizations.
(b) Make sure there are development officials and volunteers on hand to help with rescue efforts and first aid.
(c) Tell the development officer to stay in contact with the main office by wireless means, to report any losses or damages as quickly as possible, and to keep everyone updated on the situation.
(d) Union Group leaders should advise the Development Officer to forward loss and damage reports received from the field to the Regional/Central Office through wireless communication and regular contact.

Rehabilitation Stage

i. Send a main report of loss and damage in your region to the Development Officer from every Union leader.
ii. Send out the field office with specific instructions.
iii. Supporting health-related vaccination and preventative drug programs.
iv. In order to take part in the necessary rehabilitation programs.
v. With the hope of facilitating the coordination of non-governmental organizations' efforts (NGOs).

Field-Level CPP

Risk Reduction

(a) Appoint a single representative to serve as the committee's "Focal Point" and attend all relevant sessions.

(b) Take part in the Directorate of relief and rehabilitation's process of mapping local catastrophe risks and vulnerabilities and preparing risk reduction action plans.

Emergency Response

Normal Times

(a) Organize simulated exercises constantly in disaster zones and check the progress of preparation drill performed by the public between April and September every year.
(b) Pick volunteers have them undergo CPP-mandated training by April, and set up family evacuation clusters for use in case of emergency.
(c) Make physical inspection of equipment to enable given to the leader of volunteers and their condition and execute their repairs/replacement, if required.
(d) Maintaining contact between the CPP's central office and the Upazila and Union offices requires regular examination of the wireless infrastructure.
(e) Help the locals get more familiar with the cyclone preparation program and the various warning signs.
(f) To designate and maintain refuge areas, killas, and safe elevated terrain, and to inform the population of the evacuation strategy.
(g) Make the community and the volunteers aware of the hazards of opposition to disaster preparation operations, in conjunction with Upazila and Union Parishad administration.

Alert Stage

(a) Establish a Command Post and keep in touch with the Upazila, the Union Offices, and the CPP Central Office.
(b) Collect data on cyclones and keep in constant contact with the Meteorological Department and other relevant agencies.
(c) Maintain local offices with unique weather reports from CPP Headquarters.
(d) Send out an alert to the district/Upazila/union DMC chairperson, members, religious leaders, local elites, teachers, and those affiliated to other organizations.
(e) Encourage the CPP volunteers to tune in to radio broadcasts so that the Union group leaders and their coworkers may get started on the tasks outlined in the CPP brochure.

Warning Stage

(a) Ask the District Commissioner, the Upazila Administrator, and the Chairman of the District Management Committee to convene immediately (s).
(b) Help put into action the resolutions reached at the hastily called meeting.
(c) Assign CPP members to supervise the safe relocation of cattle, poultry, and other household animals to elevated terrain, killas.
(d) You should tell every field office to share the special weather bulletins with every office it manages.
(e) Share the news about the impending calamity with the public.

(f) Assist the public in evacuating to a safe location and advising them to do so if an order is given to do so.
(g) Use a megaphone, signal lights, and flashing lights to give the citizens one last warning.
(h) Report the catastrophe situation at the local level to higher authorities such as the Upazila and District administrations, the CPP Central Headquarters, the DMB, and other interested parties.
(i) Help other organizations and agencies carry out the plans they've set out.

Disaster Stage

(a) Keep Wi-Fi on and in contact with CPP Central Office, and get that loss/damage statement to the proper authorities as quickly as possible.
(b) Participate in rescue efforts and provide first aid in accordance with the requirements of the local Union and Upazila governments.
(c) Help the government in the area distribute the aid.

Rehabilitation Stage

(a) Collect information on cyclone-related losses, compile a report, and submit it to the CPP Central Office, the Union DMC, the Upazila, and the District.
(b) Help the authorities dispose of deceased people and animals.
(c) Help with health initiatives like immunization drives.
(d) Take part in the rehabilitation initiative with NGOs and other groups.

Directorate of Food (DG Food)

Maintaining sufficient food grain reserves in disaster zones is a major responsibility of DG Food. The DG is responsible for overseeing DG Food's performance of the following responsibilities in addition to the directorate's regular operations.

Risk Reduction

(a) The government's strategy on ensuring a sufficient supply of food should take into account efforts to reduce the likelihood of catastrophic events.
(b) When estimating food supply needs, it's important to factor in both potential and actual disasters.
(c) Assess the central, district, and outlying offices' susceptibility to disaster by conducting vulnerability and risk assessments of food infrastructure facilities and then taking appropriate action, such as retrofitting, as well as creating a contingency plan. When planning for new storage facilities and infrastructures, keep in mind potential threats and BNBC.
(d) Make sure the storage areas are regularly serviced so that problems don't arise.
(e) Put out the required directives and instructions to the appropriate divisions to guarantee the DG Food's risk reduction strategies are put into action.

Emergency Response

Normal Times

(a) Choose a Control Room's Focal Point and set up procedures for handling crisis situations.
(b) Choose one officer to serve as the Liaison and maintain contact with the DM&RD's Control Center.
(c) Put the owners' names on the list of vehicles like trucks and water transports and country boats, and put the total amount of products held in warehouses. Examine how well each subordinate department is doing with its quarterly work plan.
(d) If you live in a region prone to cyclones or floods, it is your responsibility to warn the locals well in advance and to take appropriate measures to secure and safeguard your food supply, silos, water and road infrastructure, and other essentials.
(e) Preparation for emergency food supply and distribution should be made if needed.
(f) Protect food and other items kept in food godowns by moving them to more secure locations.
(g) Check the supplies in LSDs and CSDs and make sure you have a sufficient advance store of food ingredients in disaster zones.
(h) The most readily available grains in catastrophe zones will be rice and wheat.
(i) Make sure you plan ahead with the local authority on everything from food delivery to storage.
(j) Keep the DM&RD, MoFDM updated on the status of food storage in disaster-stricken Upazilas and Districts on a consistent basis.
(k) In April and September of each year, you should make a thorough work plan and evaluate any preemptive measures you've implemented.
(l) Don't forget to keep a current list of who owns what semis, water taxis, country boats, etc. Keep track of the supplies, the warehouse, and the current number of grains kept.
(m) Ensure the safe shipping of food providers and make sure food storage is adequately protected.
(n) Transport for food grains must be coordinated in advance with the relevant Ministry/Department and local authorities.
(o) Make sure the LSDS and CSDs have appropriate food stored in advance, and check the supply there as well.
(p) If necessary, arrange for the immediate shipment, delivery, and distribution of food.
(q) Food and other commodities stored in warehouses should be given proper protection, and if necessary, they should be moved to more secure locations.

Alert and Warning Stage

(a) Turn on the Directorate's Control Panel.
(b) You should appoint a Liaison Officer to maintain communication with the MoFDM's EOC.
(c) Send out a warning to the relevant authorities in the regions where a catastrophe is most likely to occur.

(d) Maintain a reasonable price for grains used in cooking.

Disaster Stage

(a) Maintain constant contact with the MoFDM's EOC and run your own control room for a full day.
(b) Reports of impacted regions should be gathered using internal sources and sent daily to the IMDMCC through the DMB.
(c) Make sure there is enough food for the people living in the impacted regions by running a specific rationing system and open market sales if necessary.
(d) Avoid price manipulation and stockpiling by taking preventative measures to keep commodity costs low.
(e) Help the local government in any way you can with evacuation, rescue, and aid efforts.
(f) Under the direction of the MoFDM or NDMC, make arrangements for the rapid delivery of food supplies in accordance with the Delivery Order of DC/UNO.
(g) Find out how much food and food storage facilities were lost or damaged, calculate the costs, and make preparations for a speedy rebuild.
(h) Put together a report detailing the impacted regions' food supply and distribution every day, and forward it on to the MoFDM's Emergency Operations Center (EOC).
(i) Start operating special rationing and open market sales of food supplies with government clearance to guarantee a steady supply of food.
(j) Protect consumers against profiteering and hoarding, and maintain a reasonable market price for food by enforcing legal penalties against them.
(k) If you can help in any way, please do so to aid the local government in their efforts to evacuate people and bring in aid.
(l) Inquire about the loss or damage of food supplies and the warehouse, and then, after measuring the loss or damage, provide estimates for their rapid repair and rebuilding. You may then send them to the MoFDM to apply for grants from the NDMC and IMDMCC.

Rehabilitation Stage

(a) Set up a speedy evaluation of the damage, needs, and losses, and act accordingly.
(b) Arrange for suitable plans to be prepared and funding to be allocated for the purpose of repairing and rebuilding storage facilities.
(c) Maintain a steady supply of food staples by regularly announcing production, inventory levels, and supply guarantees.
(d) Set up a location to store relief supplies until they can be distributed.
(e) Maintain coordinated efforts between MoFDM, other Ministries, and local government to assist in the rehabilitation of impacted individuals until normality is restored.
(f) Get grain to the afflicted regions as soon as possible. Warehouse renovation and rebuilding work must begin promptly under established protocol.
(g) Release food stocks immediately per DC/UNO delivery order or after consulting MOFDM or NDMC.

(h) Prepare storage areas for the receiving and distribution of aid supplies.
(i) Cooperate with the Ministry of Foreign Disaster Management (MoFDM), other Ministries, local councils, and local government to continue making every effort to help impacted people recover until life returns to normal.
(j) Amplify your food distribution plan.

Field Offices of the Directorate of Food

In addition to their regular tasks, the District and Upazila offices of the Directorate of Food, as well as the LSDs, CSDs, and Silos in those regions, will be responsible for the following.

Risk Reduction

(a) Consider the existing and potential future catastrophe risks when estimating the area's food consumption.
(b) Connect with merchants of grain products and establish a network.
(c) Purchase grains from the local market on a regular basis to keep food reserves above a certain minimum.

Emergency Response

Normal Times

(a) Maintain communication with the local Disaster Management and Relief Coordination Committee by setting up and running an Information Centre at the workplace.
(b) One officer should be appointed as the Liaison to maintain communication with the regional DMCs.
(c) At least every three months, check in on the plans that have been put in place to ensure the safety of food supplies, warehouses, installations, transportation, and equipment by lower-level offices, the CPP, food grain dealers, and local governments.
(d) Keep a running tally of all the vehicles in the warehouse, including trucks, water transports, country boats, etc., along with their owners' identities, drivers' names, storage capacities, and other pertinent details (s).
(e) Safeguarding food storage facilities, food supplies, food and water transfers, etc., is essential in disaster zones.
(f) Check the food supply chain, logistics, and distribution system every three months.
(g) If necessary, make plans to relocate stored food and other goods to more secure locations.
(h) Inspect the stock in LSDs and CSDs and make sure there is enough food stored in secure locations in disaster zones.
(i) When a natural or man-made calamity strikes, rice and wheat will be the two most readily available grains.
(j) Store, protect, and convey food in coordination with local authorities.

Disaster Stage

(a) Maintain contact with the MoFDM's EOC and the local DMC.
(b) Appoint one person to serve as the Committee's Liaison Officer in the Control Room.
(c) Help the local government with any rescue, evacuation, or relief efforts they may need.
(d) Report the stock level and any stock damage every day to the supervising authority or the DG of the Food Directorate.
(e) Prepare for the beginning of government-ordered special rationing and open market sales of rice and wheat, and maintain a constant supply of food items.
(f) Try to maintain a reasonable market price by taking legal action against stockpiling and profiteering.
(g) Help assist the local government and council as much as you can with their efforts to get people out of harm's way and get them back on their feet.
(h) Prepare plans for the repair and rebuilding of damaged food storage facilities and submit recommendations for authorization of funds for execution, then conduct a survey to ascertain the degree of loss and damage to food storage facilities.

Rehabilitation Stage

(a) Take the necessary steps to promptly evaluate the damage and determine the extent of your losses so that you can begin making repairs and recovering as soon as possible.
(b) Get moving quickly on fixing up and rebuilding the storage areas as the money comes in.
(c) Make sure there's a steady supply of food.
(d) Facilitate the speedy distribution of food supplies in response to delivery instructions from the DC/UNO or in agreement with the MoFDM.
(e) Clear some room for the temporary storage and distribution of aid supplies.
(f) Help the local government out with their relief and recovery efforts.
(g) Increase efforts to provide food to underserved communities.
(h) Supply and distribution updates should be sent on a regular basis.

4.4.3 Ministry of Home Affairs

Search and rescue, security, relief operations, and disaster information management are all crucial responsibilities of the police, the Ansar, the Village Defence Party, the BGB, and the Bangladesh Coast Guards. The Ministry of Home Affairs is responsible for these Forces, and DCs or UNOs often oversee their deployment and operations in impacted regions.

Typically, these Forces will aid the local government in developing Action Plans, preparing for emergencies, and creating local standing orders. The Ministry will monitor these departments to make sure they're doing their part to get the job done.

The following disaster-related matters pertaining to government agencies and armed forces within its jurisdiction will be routinely reviewed by the Ministry:

(a) Preparedness Action Plan.
(b) Supplies, vehicles, and machinery used in emergency situations.
(c) Infrastructure relating to communication in locations at risk of natural disasters.
(d) Education is the art of handling catastrophic situations.
(e) Emergency practice for when disaster strikes.
(f) The drafting of the required laws

The Secretary will carry out the following in addition to his or her usual responsibilities related to responding to emergency circumstances and enforcing Action Plans developed specifically for such occasions:

Risk Reduction

(a) Carry out a thorough Ministry-wide sectoral risk assessment, and use that data to inform the development of a strategy for mitigating risks and ensuring readiness in the face of them.
(b) Make that the Ministry's strategy has enough funding to be implemented.
(c) The most efficient use of funds requires a system of checks and balances to be set up.
(d) Educate and raise awareness among your workforce.
(e) Get the Ministry to set up a framework for communicating risks across different industries.
(f) Organize its risk management and mitigation efforts into a sectoral and organizational contingency plan.
(g) Prepare the agency for potential calamities caused by earthquakes by creating a contingency plan.
(h) Arrange SAR training for the BGB, police, Ansar, VDP, and BFS&CD.

Emergency Response

Normal Times

(a) Select one Ministry Liaison Officer to act as the primary point of contact and relay that information to the relevant parties, including the DMB.
(b) One MOFDM Liaison Officer should be assigned to the DMB Control Room.
(c) Protect wireless equipment, transportation, products, and infrastructure in the affected region by coordinating the efforts of relevant Ministry-controlled entities.
(d) Make sure that the police, the Ansar, the village defense party, the Coast Guard, the fire department, and the civil defense all work together to warn the public, evacuate the area, rescue victims, and restore law and order in the aftermath of a catastrophe.
(e) Maintain Law and Order in Disaster-Prone Areas During Early Warning, Recovery, and Post-Disaster Periods.

(f) Make sure there's enough money in the emergency fund to carry out NDMC and MoFDM's requests.
(g) Plan disaster management training for BDR, police, Coast Guard, Ansar, VDP, fire brigade, and civil defense personnel in high-risk locations.

Alert and Warning Stage

(a) Make that the Ministry's subordinate agencies and armed forces are prepared to respond quickly and effectively in dangerous situations by implementing their action plans during the alert and warning phases.

Disaster Stage

(b) Manage the Ministry's Operations Center.
(c) Name one officer to serve as the EOC's primary point of contact at the MoFDM's EOC.
(d) Compile data on the current state of law and order and offer appropriate directives.
(e) Make sure the BDR, Coast Guard, Ansar, and Village Defense Party can quickly mobilize in case of an emergency (remain standby and move as per government instructions).
(f) Rescue and evacuation of people, animals, and birds from shelter/safe places, killas, and high-rise lands require collaboration and support from local government and relevant BDR, Police, Ansar, VDP, Fire Brigade, and Civil Defence.
(g) Maintain a fully functional Emergency Operations Center (EOC) at the MoFDM and ensure its continuous functioning at all times by coordinating with the BDR, Police, Ansar, VDP, Fire Service, and Civil Defense Headquarters.
(h) Maintain field level forces activity and communicate with local government and the MoFDM.
(i) Send frequent updates to the NDMC/IMDMCC and MoFDM with loss and damage data collected from regional BDR, Police, Ansar, Coast Guard, and Fire Brigade headquarters.
(j) Manage BDR, Police, Ansar, VDP, Coast Guards, etc., operations on the ground.

Rehabilitation Stage

(a) Assess the quality of the service provided, find areas of improvement, and implement them.
(b) Quickly remove BDR, Ansar, and VDP when they have finished their disaster-related tasks.
(c) After the Police, BDR, Ansar, VDP, Fire Service, and Civil Defense have completed their rescue, relief, and restoration efforts, a full report should be sent to the NDMC/IMDMCC and MOFDM.

Bangladesh Rifles (BDR)

When tragedy strikes in remote parts of Bangladesh, the civil administration, which includes the Bangladesh Rifles, may play a crucial role in preventing more damage and saving lives. When asked for help, they must come to the assistance of the local civil authority. However, in remote places where civil administration has challenges, the relevant BDR sectors will have to deal with catastrophe on their own.

All parties involved in disaster-stricken regions will be able to get their operations rolling at the right moment, according to the Bangladesh Rifles' Contingency Action Plan. Following government orders, personnel of the Bangladesh Rifles will carry out their duties in accordance with the agency or department given responsibility for the mission.

Risk Reduction

(a) Create the Bangladesh Rifles' risk management and readiness strategy by doing a thorough analysis of each sector's vulnerabilities.
(b) Create a formal sector risk assessment and then a plan for mitigation and preparation based on that evaluation.
(c) Create a plan for increasing people's ability to mitigate danger and put it into action.
(d) Establish a procedure for dealing with seismic emergencies at the agency level.
(e) Help DMB and GSB create maps showing where earthquakes are most likely to occur.

Emergency Response

Normal Times

(a) At the Bangladesh Rifles' Command Center, one Liaison Office will serve as the primary point of contact.
(b) Prepare a small group of people to respond to and recover from disasters.
(c) Make sure you have a list of hospitals, food storage facilities, and other areas where you may seek refuge.
(d) It is important for organizations to work together to ensure that their wireless communication arrangements in disaster-prone regions are coordinated with the resources and responsibilities of other organizations.
(e) Make sure you're set up to get alerts and warnings from BMD and other sources in the event of a crisis.
(f) All members of the Force need to be familiar with the various warning signs.
(g) Motivate the neighborhood to pay attention to weather and flood reports and adjust their schedules appropriately.
(h) In places prone to cyclones and floods, it is important to review the degree of readiness on all levels and to hold drills in which these measures may be practiced in conjunction with other agencies. Take precautions to safeguard their own lives, property, equipment, installations, modes of transportation (water and land), and other assets.
(i) Please provide BOPs in cyclone-affected regions with extra wireless sets.

Alert and Warning Stage

(a) On the basis of warning/alert signals provided by BD/FFWC and broadcast/telecast by Radio/TV, members of the BDR shall issue warning notice via the Upazila administration to the local population in regions likely to be impacted.
(b) Create groups to aid in rescue and relief efforts.
(c) The impacted population will get the transportation, gasoline, medication, and relief supplies they need with the help of the local BDR sectors, who will work with the local civic authorities.
(d) It will be up to individual sectors to make contact with local authorities and determine whether or not to evacuate their employees and any valuables.
(e) Each squad will have one officer overseeing their activities.
(f) Be in constant contact with BDR Central Control (24 h).
(g) It's best to be prepared for anything, so do all you can to be ready.
(h) Locate as many Force members and resources as you can.
(i) Maintain Law and Order in tight coordination with police according to work allocation by setting up a Control Room at BDR HQ and assigning a Liaison Officer to help local administration with rescue, relief, and restoration efforts.
(j) Cooperate with local government to use wireless facility to contact local DMC or MOFDM, as appropriate.
(k) Don't forget to sound the alarm at the BOP.

Disaster Stage

(a) Damage should be assessed and the local government, DMB, and BDR Headquarters notified.
(b) Move hurt people to designated areas.
(c) Find the corpses and make the necessary funeral arrangements.
(d) You need to get your hands on some important valuables.
(e) Ensure that the appropriate teams are sent to their designated locations.
(f) Make sure the appropriate teams are doing their responsibilities.
(g) Get hurt people basic medical attention.
(h) Help the government in the area with their evacuation efforts.
(i) Those who have been impacted should be provided with basic necessities such as food, water, and clothes.
(j) To the Control Room of the Bangladesh Rifles, please relay as much information as possible on the catastrophe areas.
(k) Report daily developments to BDR HQ.
(l) Make sure you have a strong backup squad ready to go in case you need to send in reinforcements.
(m) Contribute to the efforts of civic authorities to restore order in the impacted regions.
(n) If more work is required, orders might be issued.
(o) Bangladesh Rifles Headquarters to EOC at Ministry of Food and Catastrophe Management, with a copy to the Ministry of Home Affairs, about disaster information.
(p) Send out a BOP alert.

(q) Make plans to maintain the highest number of teams possible on the field.
(r) Keep the BOP battalion in touch with the Sector Headquarters through wireless means.

Rehabilitation Stage

(a) Assist the locals in building or repairing dwellings as soon as possible so they may find safe haven.
(b) Participate in the provision of aid to impacted populations by assisting local authorities.
(c) Assist and collaborate with mass immunization campaigns in impacted regions.
(d) Facilitate the delivery of aid supplies to those in need.
(e) Prepare a comprehensive report on relief and rehabilitation efforts and submit it to the Ministry of Home Affairs.
(f) Normal operating practice for the regions away from coastlines.

Location

(i) Rajshahi Sector:	This sector is located in Rajshahi city.
(ii) Rangpur Sector:	This sector is located in Rangpur town.
(iii) Dinajpur Sector:	This sector is located in Dinajpur town.
(iv) Mymensingh Sector:	This sector is located in Mymensingh town.
(v) Sylhet Sector:	This sector is located in Sylhet town.
(vi) Comilla Sector:	This sector is located in Comilla town.
(vii) Rangamati Sector:	This sector is located in Rangamati town.
(viii) Dhaka Sector:	This sector is located in Dhaka town.
(ix) Kushtia Sector:	This sector is located in Kushtia town.

Duty

Increase the number of trained Bangladesh Rifles personnel and maintain them ready to help the District and related authorities in relief and rehabilitation efforts as directed by them in the event of a catastrophe.

Bangladesh Police

Due to its widespread presence, the Bangladesh Police will be responsible for the following, on top of playing a crucial part in disaster response operations everywhere in the nation.

Risk Reduction

(a) Find out which businesses are at risk by doing an inventory. Implement a backup strategy for the organization.
(b) Identifying high-risk locations for earthquakes via coordination with DMB. Instruct law enforcement to maintain readiness and take part in emergency search, rescue, and relief efforts as needed.
(c) Hold earthquake exercises annually and assess your level of readiness.
(d) Fortify the weak structures by doing retrofits.

(e) In order to set up new facilities, use the risk maps that were created during the official risk assessments.
(f) Improve police communication systems.
(g) Create a backup means of contact for usage in case of emergency.
(h) Maintain communication between governments through police radio.
(i) Help raise people's understanding of how to prepare for and respond to disasters.
(j) Make sure there's a record kept of what kinds of tools may be utilized for SAR and resuscitation efforts.

Emergency Response

Normal Times

(a) Choose one Directorate Liaison Officer to serve as the group's primary point of contact for disaster management.
(b) Police officers should be educated in emergency care procedures such as CPR and first aid, as well as evacuation, rescue, and relief efforts.
(c) Make a list of the places that are most likely to be hit by cyclones, floods, or any other kind of natural catastrophe, and make sure the police stations and other facilities in those regions know what to do to prepare for any eventuality.
(d) Bring the wireless networks of the Bangladesh Police, the Bangladesh Rifles, the Bangladesh Water Development Board, the Ministry of Foreign Affairs and Disaster Management, the Bangladesh Red Crescent Society, the Dhaka Metropolitan Board, the Bangladesh Railways, the Central Power Board, the Bangladesh International Water and Telecommunications Company, and so on, into harmony with one another.

Alert and Warning Stage

(a) Create a Disaster Command Center at the Division, Range, and District levels.
(b) One officer should be designated as the Liaison Officer and responsible for maintaining communication with the MOFDM Control Room.
(c) Protect the Bangladesh Police's VHP communication system, warn those who need to be warned, and share data with emergency management organizations.
(d) Maintain proper communication with law enforcement facilities in disaster-prone regions.
(e) Get transportation systems in working order and stored in a central location so they may be quickly deployed to disaster zones.
(f) Cooperate with local government and local DMC for any efforts to be taken to battle any crisis, and maintain direct touch with various levels of authorities such as the Divisional Commissioner, DC, UNO, Chairman, and others.
(g) If any authority figure should demand it, tell everyone involved to prioritize wireless communications relevant to the crisis.

Disaster Stage

(a) As soon as evacuation orders are given, prepare the threatened population for evacuation in close coordination with local people, agencies, and local government.
(b) Maintain effective law and order and the protection of the evacuees by sending in backup police forces to the affected regions as quickly as feasible.
(c) Keep a tight eye out for any illegal or antistate activities, and respond accordingly.
(d) It is essential to quickly acquire cars for cleaning up catastrophe debris and other tasks.
(e) Maintain internal and exterior traffic management and law and order in disaster-stricken regions.

Rehabilitation Stage

(a) Arrange security of government property and installations damaged in the disaster.
(b) Control traffic in regions near damaged roads, bridges, etc., if required.
(c) Help the government or other local organizations bury the deceased and bury those who have been rescued when they have been found.
(d) Help those who have been hurt or afflicted by a calamity by providing first aid.
(e) Help the government set up emergency shelters if required.
(f) Help the government in the area prevent the misappropriation of aid funds and supplies.
(g) Help with the cleanup and restoration efforts until life can return to normal.

Directorate of Ansar and VDP

The widespread presence of Ansar and the Village Defence Party (VDP) around the nation gives them a unique opportunity to aid in catastrophe preparedness and recovery. Publicizing warning signs, rescuing victims, evacuating them to safety, ensuring law and order are maintained, and rehabilitating damaged areas are all part of their disaster response.

Risk Reduction

(a) Conducting a thorough review of sectoral risks can help you create a strategy to deal with them.
(b) Set aside funds in the budget to carry out the strategy.
(c) The most efficient use of funds requires a system of checks and balances to be set up.
(d) In order to prepare employees to volunteer during times of catastrophe, such as earthquakes and fires, education and awareness training must be undertaken.
(e) Get the Ministry to set up a framework for communicating risks across different industries.
(f) Create a backup strategy for its risk management and mitigation procedures. Raise everyone's consciousness about the need of being earthquake ready.
(g) Gather into the following teams to better handle earthquake emergencies:

i. Rescue and Search Committee
ii. The Junk Removal Board
iii. Treatment of Emergencies
iv. Disaster aid and evacuation planning

(h) Communication with DMB to plan earthquake exercises and assess the current state of preparation.
(i) Work in tandem with DMB to map out the regions of their service territories that are most at risk from earthquakes.

Emergency Response

Normal Times

(a) Provide Ansar and VDP personnel with training on how to carry out their roles in sending out a "Warning/Alert signal," evacuating victims, rescuing them, administering first aid, distributing supplies, and rehabilitating
(b) Platoons will be formed as follows inside the Ansar and VDP Company to carry out disaster relief operations.

i. Shifting and Rescue Platoon
ii. Relief Platoon
iii. Reconstruction Platoon
iv. First aid Platoon

(c) Get the company/platoons mentioned in (b) above some basic and refresher training.
(d) Preparatory drills for cyclones should be held in all cyclone-prone regions every year in April and September under the directives of DC or the DMC.
(e) Promptly issue orders to all relevant Ansar and VDP units to inform them of the above SOD.
(f) Keep in touch with the DMB, DC, CPP, BRCS, and UNO/Chairman to coordinate the tasks that need Ansar and VDP's involvement.
(g) Locate safe areas for people and cattle, and maintain communication with the CPP, volunteers, and locals to ensure that responsibilities are distributed fairly.
(h) The coastal and island populations should be won over by publicizing cyclone warning signals.
(i) Help the Upazila and Union Parishad keep their storm shelters and earthen mounds in working order.
(j) Ansars should always be prepared to evacuate based on orders from Bangladesh Betar or any other authority working with local DMC.
(k) Disaster preparation programs for battling cyclones and floods should be organized by the District and Assistant Adjutant of Ansars in close cooperation with the District and Upazila Police, the Red Crescent, the Fire Services, and Civil Defence.

(l) The District Adjutant of Ansars is responsible for making sure that all authorities are aware with the environment, communications, etc., of the catastrophe zones and that they visit such areas as often as feasible.

Alert and Warning Stage

(a) Get everyone together in the Directorate's Control Room and brief them on the situation.
(b) Make sure that warnings go out to the public as soon as feasible.
(c) Those at Ansar and VDP should be warned.
(d) Communicate regularly with the District of Columbia, the United Nations Office, the Chairman of the Union Parishad, and any other relevant government offices or agencies to ensure smooth cooperation.

Disaster Stage

For Cyclone

(a) If feasible, go door-to-door in cyclone-at-risk regions and get the word out as soon as possible.
(b) Follow orders from authorities to move at-risk individuals to a safer location.
(c) Safeguard the homes of those who have been evacuated and take all necessary precautions.
(d) Facilitate worship and maintain order in storm bunkers and other evacuation centers as directed by the proper authorities.

For Floods

(a) Upon receiving orders from the DC/UNO, all officials in disaster-stricken regions must immediately contact the local administration for instructions and go to the scene of the incident.
(b) Under the direction of an officer or platoon leader, local Ansar platoons will conduct rescue operations and the burial of people and animal carcasses.
(c) As compassionately as possible, provide aid to those who need it. Keep an eye on the property of those who had to leave their homes during an evacuation.
(d) Those who have received training in epidemic vaccination should begin working with the Health Department on a vaccine campaign immediately.
(e) Create a report on the flood's effects and the epidemic's spread.
(f) Coordinate with law enforcement to take the necessary measures to bring crime levels in the impacted regions under control.

Rehabilitation Stage

(a) Gather in the affected region and report to the local authorities for duty.
(b) Help the government get stranded, distraught individuals to emergency shelters if they need it.
(c) Help the local government maintain order at designated aid shelters.
(d) Help the locals and authorities treat those who are suffering and provide aid.

(e) Help bury the deceased and dispose of animal carcasses to keep health and sanitation from deteriorating in impacted communities.
(f) Help those in need by rebuilding their homes or taking on other projects that promote a spirit of cooperation and self-sufficiency.
(g) Help the authorities put together an accurate account of the damage and losses.
(h) Be a part of any government or non-government long-term rehabilitation program, including agri-rehabilitation.

Fire Service and Civil Defence (FSCD)

In addition to its normal functions, the department will undertake the following.

Risk Reduction

(a) Determine the state of FSCD service provision in the nation and its projected growth.
(b) Find suitable spots to set up the Fire Brigade Headquarters.
(c) Create a strategy with an approximate budget to open the necessary number of Fire Brigade Offices.
(d) Invest more resources into the Fire Brigade Office's staff and apparatus.
(e) Create an education campaign to teach the general public about fire prevention, safe evacuation, and first aid.
(f) You should fortify the building in light of the possibility of an earthquake or fire, as well as the Fire Brigade's other buildings.
(g) Every six months, you should revise and update your agency's emergency plan.
(h) Coordinate with the Geological Survey Team to acquire earthquake risk maps and develop mitigation strategies for disaster-prone locations.
(i) Make a list of everything you'll need for search-and-rescue efforts, and make sure it's up to date.
(j) Perform regular exercises on earthquake preparation and disaster management, and evaluate your level of readiness.
(k) Make sure you have an up-to-date database listing all the groups that donate blood and organs.
(l) Prepare cities for earthquakes and fires by training urban volunteers. They should get regular training in tandem with DM&RD, DMB, and City businesses.

Emergency Response

Normal Times

(a) One Liaison Officer should be appointed as the main point of contact; everyone else should be made aware of this.
(b) To better deal with calamities like cyclones, floods, fires, and earthquakes, it is important to contact organizations focused on social welfare and education in order to recruit volunteers and provide them with the necessary training.
(c) In normal, disaster, and rehabilitation phases, volunteers will be trained (in coordination with CPP, BDRCS, local administration, Police Directorate, and

all concerned organizations) to implement a coordinated Action plan for the protection of assets and food storage facilities, the evacuation and rescue of people and livestock, and relief and rehabilitation-related work.
(d) Create an updated list of trained individuals' contact details.
(e) Gather supplies for the fire department, the volunteers, and the relief effort, and keep them at the fire station.
(f) Improve the efficiency of the Fire Department's radio network.

Alert and Warning Stage

(a) Once the alarm is received, all Fire Service and Civil Defense units should be notified and kept on readiness.
(b) Together with the District/Upazila administration and related local agencies, organize mock drills for fire extinguishing, rescue, evacuation, and transportation of wounded individuals.
(c) Keep in touch with the local government, police, BDRCS, and other welfare organizations to coordinate efforts.
(d) Move vulnerable individuals and resources (livestock) to shelters or other secure locations.

Disaster Stage

(a) In order to comply with the directions of the Division Commissioner, DC, UNO, and the concerned DMC, it is necessary to set up a control room at the headquarters, division, and district levels.
(b) As soon as a catastrophe strikes, the local Fire Service and Civil Defence units in the affected District and Upazila should get in touch with the local government to get instructions on how to proceed with disaster relief efforts.
(c) Everyone who wants to help out with Fire Service and Civil Defense should meet up at the closest station.
(d) Firefighters and other emergency responders must rush to the scene as soon as possible to help with fire suppression, search and rescue, first aid, evacuation, transportation of wounded individuals, pumping water out of food storage facilities, etc., as directed by local authorities.
(e) If it becomes necessary, the highest-ranking officials will bring in more personnel from less afflicted regions to bolster the service.
(f) The senior officer's responsibilities include overseeing the efforts of his service and coordinating with other agencies providing aid to the impacted regions. Every division under his Directorate's purview will work together to support him.
(g) Ensure that the local control room is updated on any and all work that is being done.
(h) Get the volunteers involved and put them to work doing what they're best at.
(i) Keep up your efforts to combat antisocial behavior and provide assistance to law enforcement.

Rehabilitation Stage

Workers from the Fire Service and Civil Defence will take charge of post-disaster aid and rehabilitation with the help of the local government and other social organizations after emergency activities have concluded.

(a) Help those in need by giving them food, housing, and medical attention.
(b) Find lost or stolen government or individual property.
(c) Fix the messed up homes.
(d) To eliminate potential danger, demolish any structures that aren't up to code.
(e) Bring in clean water for consumption.
(f) Union Parishad and/or Anjuman-e-Mafidul Hasan and/or Anjuman-e-Mufidul Islam and/or any comparable organization should be contacted to remove trash and bury or remove dead corpses and animal carcasses.
(g) Vaccinate the masses.
(h) Provide aid and help with recovery for those impacted.
(i) Find missing folks and get them back to their loved ones.
(j) In other words, set up a hub for information.

4.4.4 Bangladesh Coast Guards

In addition to its normal functions, the department will undertake the following.

Risk Reduction

(a) Bangladesh Coast Guards' existing and future service requirements may be determined by conducting a national catastrophe risk assessment.
(b) Locate suitable areas in which to set up the offices.
(c) Create a plan of action, including a rough budget, to open the right number of offices.
(d) Increase the personnel and resources available to Coast Guard offices.
(e) Create a program to raise people's awareness of the Coast Guard's role in emergency situations in order to enhance their services, particularly in the areas of safe evacuation and first aid.

Emergency Response

Normal Times

(a) One Liaison Officer should be appointed as the point person, and everyone should be made aware of this.
(b) Create an updated list of trained individuals' contact details.
(c) Get all the BCG gear and the relief supplies to the offices.
(d) Refine the Radio System for Better Coverage.

Alert and Warning Stage

(a) As soon as the BCG receives the alarm signal, they are to be put on standby.
(b) Coordinated Action Plans should be practiced with the District/Upazila administration and relevant local agencies for firefighting, search and rescue, evacuation, and transportation of wounded individuals.
(c) Keep in touch with the local government, police, BDRCS, and other welfare organizations to coordinate efforts.
(d) Move those in danger and their assets (livestock) to shelters or other secure locations.

Disaster Stage

(a) In order to get things rolling, you need to set up a command center at headquarters and field offices, get in touch with the division commissioner, the DC, the UNO, and the relevant DMC leaders, and start working.
(b) Local BCG units in affected districts and Upazilas should get in touch with their respective governments as soon as possible when a catastrophe strikes in order to get instructions on how to proceed with relief efforts.
(c) Ensure that the local control room is updated on any and all work that is being done.
(d) Get the volunteers involved and put them to work doing what they're best at.
(e) Keep up your efforts to combat antisocial behavior and provide assistance to law enforcement.
(f) Assure offshore island and sea rescue operations and emergency medical care and report back to DMB.

Rehabilitation Stage

(a) Workers from the BCG, aided by local authorities and other relief organizations, will take on the following post-disaster relief and rehabilitation tasks after the emergency situation has been stabilized.
(b) Help those in need by giving them food, housing, and medical attention.
(c) Find lost or stolen government or individual property.
(d) Fix the messed up homes.
(e) Bring in clean water for consumption.
(f) Helpful DMC will assist with trash collection and the disposal of corpses and animal remains.
(g) Provide aid and help with recovery for those impacted.
(h) Find missing folks and get them back to their loved ones.
(i) In other words, set up a hub for information.

Ministry of Defence

In addition to its normal functions, the Ministry will undertake the following.

Risk Reduction

(a) Complete a comprehensive risk analysis of defense concerns relating to natural and man-made catastrophes and hazards.

(b) Create a Ministry of Defense plan for preparing for and mitigating risks.
(c) Protect your people, your property, and your resources by taking preventative measures in dangerous regions.
(d) Take on planning for the Armed Forces' usage in assistance to the civil authorities in concert with the DM&RD, MoFDM based on request and necessity.
(e) Before the cyclone and flood seasons begin, make sure proper preparations have been made to ensure the safety of the residents, workers, property, transportation, etc., in the affected regions.
(f) Conduct seismic preparation training for all employees and draft a policy.
(g) Set up earthquake exercises and practice emergency response procedures.
(h) Create a comprehensive strategy for securing the impacted population and locating missing persons, as well as deploying search-and-rescue teams and distributing aid.
(i) Establish a plan to identify earthquake-prone sites and estimate potential losses.

Emergency Response

Normal Times

(a) Appoint a reliable someone to serve as the Disaster Management Focal Point, who will be in charge of communicating with relevant authorities and keeping everyone informed.
(b) Before the season begins, make sure that the necessary and adequate preparations have been made, together with the requisite equipments, to extend all out supports for the security, evacuation, and rescue purposes of the people, particularly in disaster zones.
(c) Maintain efficient warning signal distribution, evacuation, rescue, and rehabilitation assistance for civil authorities.
(d) Create a framework for responding to emergencies in certain industries.
(e) Make a Ministry emergency plan.
(f) Adhere to the NDMC's decision and provide funds for disaster relief efforts.
(g) Help civil authorities with warning signal transmission, evacuation, rescue, and rehabilitation by ensuring efficient coordination among line organizations.
(h) Work in tandem with the Ministry of Defense and Military Affairs (MoFDM) to ensure that the Armed Forces are called upon to assist civil authorities in accordance with the Ministry's stated needs.

Alert and Warning Stage

(a) Maintain constant Control Room operation (24 h).
(b) Appoint one officer to serve as the EOC's liaison at the MoFDM.
(c) Gather disaster data from BMD and FFWC, and share it with relevant parties, including the MoFDM.

Disaster Stage

(a) It is essential to have constant communication with the MoFDM's EOC and the armed forces' control rooms in order to get timely and accurate updates on the situation.
(b) Assist in keeping the MoFDM and AFD on the same page so the Field Task Force may function efficiently.

Rehabilitation Stage

(a) When the NDMC and you agree it is time, end military relief efforts.
(b) Provide the NDMC/MoFDM with updates on military and civilian efforts to restore affected areas.

Bangladesh Meteorological Department (BMD)

In addition to normal duties, the Bangladesh Meteorological Department will perform the following duties.

Risk Reduction

(a) Create a plan for the department to deal with risks in certain sectors and be ready for them.
(b) Make sure money is set out in the budget for programs and activities that mitigate risk.
(c) Improve cyclone prediction and alert methods on a regular basis.
(d) Consistently disseminate all relevant data and ensure that cyclone prediction systems and procedures are continually improved.
(e) Those print and electronic media, as well as the MoFDM, need to have better equipment capabilities for the fastest transmission of the information/warnings to all involved, such as fax and email arrangements.

Emergency Response

Normal Times

(a) Create a framework for responding to emergencies in certain industries.
(b) Create a backup plan.
(c) Maintain a vigilance in monitoring the weather and working to improve cyclone prediction processes and information dissemination.
(d) Maintain the highest possible level of efficiency in the fastest means of communicating weather warnings to everyone who needs to know about them. The SWC of BMD must make arrangements with Radio, Television, and the MoFDM for the transmission of messages by fax, email, and other means.

Alert Stage

(a) Cyclone warnings should be sent as soon as feasible, preferably 36 h before a depression forms in the Bay of Bengal.
(b) Send CPP updates on the development of the depression in the Bay of Bengal through telegraph, telephone, or teleprinter so that it may take the necessary measures, including the transmission of information to all parties involved.

(c) Warning signals code "Whirlwind" per Annexure-A should be sent through phone, teleprinter, telegraph, fax, email, etc. to all relevant authorities.
(d) For the benefit of the general public, please prepare and submit a Special Weather Bulletin for broadcast/publicity via national news media, including all stations of Radio and Television and national publications. Coordinate publicity outside of usual broadcasting hours with the SWC of the BMD, Bangladesh Betar, and Bangladesh Television in the event of Local Cautionary Signal no. 3.
(e) EOC at the MoFDM, the DRR, the CPP, and the BRCS should receive Special Weather Bulletins so that they may make preparations as needed.

Warning Stage

Publicize warning signals at each of the following specified stages.

(a) Warning	24 h before
(b) Danger	At least 18 h before
(c) Great Danger	At least 10 h before

The same warning signals are to be repeated to the EOC at the MoFDM, Control Room of the DMB, the DRR, the CPP, and the BDRCS.

The following information should be mentioned in the signals to be disseminated.

(a) Exactly where the eye of the storm is.
(b) A measure of the storm's speed and its predicted path.
(c) The Upazilas of the impacted Districts should be mentioned if at all feasible.
(d) Gale winds (defined as winds with speeds in excess of 32 miles/h or 51.84 km/h) should be expected during their designated times in a variety of locations.

Fax, telephone, teleprinter, or telegraph communications labeled "Hurricane" should be transmitted to the designated locations in Annexure-A in the event of an emergency. In the event of a major emergency, communications labeled "Typhoon" and delivered through Fax, telephone, teleprinter, or telegraph to the locations listed in Annexure-A should be expected. Send separate communications to the contact details provided in Annexure-A by Fax, telephone, teleprinter, or telegraph using the codename "Water ways and Authority" for Inland Water Transportation.

Bangladesh Betar and Bangladesh Television stations around the country should be notified to broadcast alerts and warnings.

Rehabilitation Stage

i. Participate in these activities alongside the DMB.
ii. Evaluate the cyclone's impact in light of the cautionary indication.
iii. Amass information from impacted regions for scientific analysis.
iv. Seek for public feedback on signal issues.

Bangladesh Space Research and Remote Sensing Organization (SPARRSO)

In addition to normal duties, the SPARRSO will perform the following duties.

Risk Reduction

(a) Make the necessary satellite maps and imagery to back up your industry's plan for mitigating risk.
(b) Make sure money is set out in the budget for programs and activities that mitigate risk.
(c) Enhance the use of images in early warning systems for natural disasters.

Emergency Response

Normal Times

(a) You should choose one of your top employees to serve as the SPARRSO's Focal Point.
(b) Provide BMD, FFWC, DAE, Agricultural Information Service, and DMB with processed data gleaned from satellite photos of natural disasters including drought, flood, cyclone, tornado, storm surge, etc.

Ministry of Water Resources

In addition to normal duties, the Ministry of Water Resources will perform the following duties.

Risk Reduction

(a) Policy, planning, and programming in the water industry should take catastrophe risk reduction into account.
(b) Make sure the Ministry of Water Resources has a plan in place to deal with any disasters and protect its citizens.
(c) Make sure money is set out in the budget for programs and activities that mitigate risk.
(d) Set up a reliable method for monitoring and assessing progress toward the goals of the risk reduction programs and activities.
(e) Set up a method of sharing information about potential dangers across industries.

Emergency Response

Normal Times

(a) Appoint one of your senior staff members to serve as the Ministry's Point of Contact.
(b) Attend the NDMC and IMDMCC meetings and provide BWDB instructions on how to put them into action.
(c) Direct its various divisions and departments to carry out the decisions made at the NDMC, IMDMCC, and NDMAC meetings.
(d) Send out orders to the District, Upazila, and Union Parishad levels to help the government, non-governmental organizations, and the general public.
(e) Integrate the Ministry's disaster management policies and procedures.
(f) Make ensuring that catastrophe prevention measures are included into national and international plans.

(g) Make sure the Ministry has enough money to fund its risk management and mitigation initiatives.
(h) Build embankments and keep them in good repair in flood-prone locations; run sluice gates; clean out and repair storm drains.
(i) Manage and run the Flood Information Centre (FIC) and the Flood Forecasting and Warning Centre (FFWC) (April-November).
(j) Keep an eye on the water levels of all of the main rivers.
(k) Report the status of flooding once a week.
(l) Implement measures to lessen the likelihood of harm, such as flood insurance, to protect against floods.
(m) Create a plan of action for the Ministry's disaster risk reduction and management initiatives that can be implemented in the event of an emergency.

Cautionary/Warning Stage

(a) Keep an eye on the water levels of all of the main rivers.
(b) Offer flood alerts and forecasts.
(c) When dealing with a crisis, it's important to keep everyone informed and updated as you work to recover.

Disaster Stage

(a) Report flooding conditions every day.
(b) You should secure the sluice gates and fix any holes, leaks, or weakened spots in the embankments.
(c) Inspect the damage as soon as possible and take corrective measures.
(d) Ensure the Ministry's warning signals are being effectively disseminated.

Rehabilitation Stage

(a) Manage the process of infrastructure repair, rebuilding, and rehabilitation, including loss and damage assessments.
(b) Restore domestic, industrial, and export projects' infrastructure, logistics, and installations. Agriculture, fishing, and industrial revitalization projects will be given first attention.

Bangladesh Water Development Board (BWDB)

As a part of its normal functions, Bangladesh Water Development Board will perform the following duties.

Risk Reduction

(a) Assist the Ministry in making sure disaster preparedness is a part of all water-related strategies and programs.
(b) Create a plan of action based on the Ministry of Water Resources' readiness and risk reduction strategy for the water sector.
(c) Carry out the plans and measures for decreasing danger.
(d) Efforts, such as better technology and better models, should be made to better anticipate floods.

(e) Invest in cutting-edge equipment and qualified staff to fortify the Flood Prediction and Warning Centre.
(f) When planning the embankments, it's important to think about potential dangers both now and in the future.
(g) Facilitating local government's ability to mitigate risk.

Emergency Response

Normal Times

(a) As Focal Point, BWDB's FFWC will work with the Centre's Deputy Director, who will serve as Liaison Officer.
(b) Perform maintenance on water discharge equipment, such as sluice gates, on finished embankment sections.
(c) Keep a close eye on the state of the embankment and fix any holes or other damage at once.
(d) Maintain FFWC operations between April and November.
(e) Create satellite command centers for early warning systems at district offices of the supervising engineer and executive engineer.
(f) Gather up the Meteorological Office special weather reports and read them on a regular basis.

Cautionary/Warning Stage

(a) Send out orders and warnings to frontline employees through the command center.
(b) Determine who will serve as the Liaison Officer and be responsible for communicating with the DMB.
(c) Maintain a system to regularly check for embankment problems such as leaks, breaks, collapses, and other forms of damage. Assemble a team of locals to help with the urgent repairs that need to be done.
(d) Get the necessary supplies available for usage in the impacted regions.
(e) Give orders to lower-level authorities to make available technical expertise, equipment products, transportation, etc., to aid local civil administration in rescue, evacuation, and relief operations.

Disaster Stage

(a) Maintain constant (24-h) Control Room operations, notify all relevant parties of an impending catastrophe, and issue orders to take protective measures to prevent loss of life, property, machinery, transportation, and other essentials.
(b) Determine what technical resources and materials will be needed to make immediate repairs to broken or destroyed infrastructure and supply lines.
(c) Inquire with the MoFDM about any and all emergency rehabilitation support needs.
(d) If there is an immediate threat to life, property, or the ability to evacuate, then appropriate measures must be taken.

Rehabilitation Stage

(a) Assess the full amount of the damage or loss and, if practicable, make plans for rehabilitation, including cost estimates, using money from your own department or seeking funding from other sources.
(b) Create appropriate programs to ensure the safety of newly constructed offshore islands.
(c) Assist the government and any other organizations involved in the recovery efforts in any way you can.

Field Offices of Chief Engineer/Superintending Engineer/Executive Engineer/Assistant Engineer (Cyclone-Related) of the BWDB

The field-level officials and staff of BWDB shall perform the following duties in their respective areas.

Risk Reduction

(a) Take the necessary steps to analyze the water industry's risks at the local level and draft a strategy to mitigate them.
(b) Build the berms in accordance with the BWDB-approved plan that takes into account the specific conditions of the site.
(c) Analyze the current state of things and keep track of information on things like embankments, polders, and sluice gates. Preserve the system's integrity by doing routine maintenance.
(d) Construct protective embankments on schedule while meeting or exceeding quality standards set by the appropriate authorities.

Emergency Response

Normal Times

(a) Chief Engineers and Supervising Engineers shall appoint Disaster Focal Points for their departments and notify the Federal Emergency Management Agency's Federal Incident Command System.
(b) Arrange for the collection of specific weather bulletins/news and tell those involved at field level. Direct for security measures to be taken in embankments and other sites.
(c) Join the neighborhood DMC and have a chat.
(d) The field's Chief Engineer will designate a Liaison Officer to serve as the point of contact for the DMC.
(e) Assist the civil administration in coordinating and carrying out rescue, evacuation, and relief activities by providing the necessary equipment, supplies, transportation, and technical expertise.
(f) Fix any damaged sluice gates and any leaks, breaks, holes, or weak spots in the embankment in their section. Also, have your repair supplies handy in a convenient location.

(g) Hire watchmen to prevent saltwater from entering the polders during a cyclone's tidal bore and to prevent the sluice gates from being damaged.
(h) Report the status of the sluice gate, embankment, and other works, as well as the repair and restoration process, at regular intervals to the appropriate authorities.

Alert and Warning Stage

(a) Protect the lives of BWDB workers by taking the necessary precautions, and keep BWDB property, equipment, transportation, etc., safe.
(b) Retain contact with the regional DMC and work together with other organizations.
(c) Give top attention to urgently needed infrastructure repairs and replacements.

Disaster Stage

(a) Maintain a 24-h-a-day, seven-day-a-week information center and assign a point person to work with the regional DMC.
(b) If there is an emergency, contact the BWDB data center and the local government's Control Room.
(c) Help assist the local government in their efforts to save lives, get people out of harm's way, and provide aid.
(d) Take care of any breakdowns, broken installations, or interrupted supplies by rounding up the necessary technical resources and personnel.
(e) Shortly after the water recedes, damage and loss must be assessed, and a plan for repair, rebuilding, and reinstallation must be initiated in line with the government's short-term and long-term plans.
(f) Tours of the affected region should be used for immediate preventative measures.
(g) In the event of a problem or unusual condition, for which your office cannot provide a solution, you should contact the local civil administration or DMC for assistance.
(h) When necessary, work with the local government and the Disaster Management Coordinator to take whatever steps are needed to ensure the safety of people and their belongings, as well as any necessary evacuations.

Rehabilitation Stage

(a) Prepare plans and drawings as soon as feasible for the repair, reinstallation, and rebuilding of physical infrastructure, embankments, and sluice gates at needed sites after conducting a loss/damage assessment.
(b) When aiding the government, it's best to work with them as much as possible.
(c) Locate potential sites where embankments may be built to defend against tidal bores or floods, make designs, and submit an application for funding to the relevant authorities.

Bangladesh Water Development Board (Flood-Related Activities)

In addition to normal functions and Contingency Plan on floods, the BWDB will also perform the following duties.

Risk Reduction

(a) To reduce risks in the industry over the long term, it is important to first analyze them.
(b) When planning the embankments, protection walls, sluice gates, and other infrastructure, it is important to take into account the possibility of any and all hazards and to learn from past disasters.
(c) Don't cut funding for emergency measures like floodwall construction, embankment repairs, and the replacement of sluice and lock gates.

Emergency Response

Normal Times

(a) Manage the FFC effectively, refine the process for making flood predictions, and then report the results to the relevant authorities.
(b) Maintain the "Flood Information Centre" each year from April through November.
(c) Starting in April of each year, have a flood information Sub-Centre set up at the field level.
(d) During the monsoon season, it is important to compile meteorological predictions and the water levels of all major rivers draining into Bangladesh and India. The BWDB will inquire with the Ministry of Water Resources about getting data from India.
(e) Maintain a consistent schedule of press releases informing the public of the latest weather developments.
(f) MoFDM and DCs on high alert.
(g) Flood updates should be sent to everyone who cares about the issue every week.
(h) Give annual April instructions to all subordinate offices on the following:
    (i) Coordination with local administration
    (ii) Supply of required implements, materials, transports, articles, and technical know-how
(i) Share the Board's Information Cell's operations with the DMB and MoFDM.
(j) See to ensuring that the IMDMCC, MOFDM, and DMB are all working together.
(k) Appoint one officer in the Board Office to serve as the liaison between the Board and the EOC of the MoFDM. This duty will fall within the purview of the FFWC's Deputy Director.

Alert and Warning Stage

(a) Due to the short notice of flash floods, it is necessary to take measures to inform everyone by telephone, telex, and wireless technology.
(b) Put watchmen on the embankments to detect any signs of leakage, breach, or holes, and to notify the appropriate authorities. Involve locals in emergency repairs as soon as possible, preferably by the end of April. The tools and supplies needed for this task must be kept in a convenient location.

(c) Protect lives, resources, inventory, and equipment by keeping the authorities on high alert.

Disaster Stage

(a) Maintain a 24-h Information Cell and Flood Control Centre, and assign a Liaison Officer to the Ministry of Food and Disaster Management's Emergency Operations Center (EOC).
(b) Notify the MoFDM's EOC and IMDMCC of any developing emergencies.
(c) Assemble the necessary technicians and supplies to fix broken infrastructure and restore power.
(d) Help the local government in any way you can during rescue, evacuation, and relief activities using their administrative machinery on the ground.
(e) BWDB's Chief Engineer, Supervising Engineer, Executive Engineer, and Assistant Engineer on the ground should be given the following directives in accordance with their delegated authority.

    (i) Participate in local DMC operations
    (ii) To assemble all necessary technical personnel and resources to rebuild damaged infrastructure and supply lines
    (iii) After assessing the loss/damage and water level decline, a strategy and schedule for repair, rebuilding, and reinstallation should be formulated in accordance with the government's short- and long-term plans as soon as feasible.

(f) You should assign managers to make regular trips to the impacted regions.
(g) If there is an issue that can't be resolved or the IMDMCC/assistance NDMC's is required, the MOFDM should be contacted.
(h) Act in accordance with the situation's urgency in order to save life and property.
(i) Send daily flood reports to the following offices:

    (1) President's Office
    (2) Prime Minister's Office
    (3) Ministry of Food and Disaster Management
    (4) Ministry of Home Affairs
    (5) Ministry of Information
    (6) Secretary, Ministry of Agriculture
    (7) Secretary, Ministry of Water Resources
    (8) Secretary, Ministry of Fisheries and Live Stock
    (9) Secretary, Ministry of Power, Energy and Mineral Resources
    (10) Secretary, Ministry of Education
    (11) Secretary, Roads, and Highways Division
    (12) Secretary, Railways Division
    (13) Secretary, Ministry of Water Transport
    (14) Secretary, Ministry of Health and Family Welfare
    (15) Secretary, Ministry of Industries

(16) Secretary, Ministry Local Government, Rural Development and Cooperatives.
(17) Secretary, Ministry of Social Welfare
(18) Secretary, Ministry of Defence
(19) DG, Disaster Management Bureau
(20) DG, Relief and Rehabilitation Directorate
(21) Director General, Radio/Television
(22) Concerned Chief Engineers
(23) Concerned Deputy Commissioners
(24) Concerned Upazila DMC leaders

Rehabilitation Stage

(a) Determine the extent of the damage and the need for repair and restoration as soon as possible.
(b) See to it that projects for domestic, industrial, and export use are back up and running as soon as feasible by restoring their infrastructure, logistics, and installations as soon as possible. Priority will be given to projects that aid agriculture, fishing, and rehabilitating of the manufacturing sector.
(c) Give aid and cooperate with the government and other organizations working on the rehabilitation effort.
(d) Make fresh schemes for flood management, damage reduction, and prevention.
(e) Consider the pros and cons of the existing strategy to help shape its future direction.

Flood Forecasting and Warning Centre (FFWC)

In addition to normal functions and Contingency Plan on floods, the will also perform the following duties:

Risk Reduction

(a) Evaluate the potential for flooding and create a strategy to protect against flooding and other water-related disasters in the future.
(b) Further investigation into methods of increasing the lead time for flood and flash flood forecasting and improving the major links to regional flood forecasting sources is needed.
(c) Seek out the places of Bangladesh most at risk for flooding.
(d) Locate the regions in Bangladesh most at risk for sudden flooding.

Alert and Warning Stage

(a) In the event of flooding or a flash flood, it is necessary to take the appropriate measures to notify all affected parties.
(b) If you want to make sure DMB and BMD are adequately prepared for flooding, you should share knowledge regarding long-, medium-, and short-term flood prevention with them.

Field-Level Offices of Chief Engineer/Superintendent Engineer/Executive Engineer/Assistant Engineer (Flood-Related) of BWDB

In addition to their normal functions, these offices will perform the following duties in light of delegation of powers in their respective spheres:

Risk Reduction

(a) Take the necessary steps to analyze the water industry's risks at the local level and draft a strategy to mitigate them.
(b) Build the berms in accordance with the BWDB-approved plan that takes into account the specific conditions of the site.
(c) Analyze the current state of things and keep track of information on things like embankments, polders, and sluice gates. Preserve the system's integrity by doing routine maintenance.
(d) Construct protective embankments on schedule while meeting or exceeding quality standards set by the appropriate authorities.
(e) Improve the reliability of the data system used for flood prediction.

Emergency Response

Normal Times

Ensure the following:

(a) Coordinate operations at the Flood Information Centre from April through November each year.
(b) Coordinate the collection of data on precipitation and river levels at various sites along rivers that begin in India.
(c) Flood forecasts should be sent to the BWDB's Flood Control Room and the Ministry of Water Resources' Control Room through their respective Heads of Office.
(d) Increases in flood levels and flash floods should be reported to the BWDB, the Ministry of Water Resources, the Emergency Operations Center of the MoFDM, as well as any concerned district commissioners or municipal emergency management officials.
(e) Share the weekly flood position with everyone who cares.
(f) See if there is a local DMC meeting you can attend.
(g) Provide the local government with the necessary mechanical tools, transportation, resources, and tactical support for rescue, shifting, and relief operations.
(h) The local Emergency Operations Center should have a designated Liaison Officer who will maintain communication with them.
(i) When a flash flood is imminent, time is of the essence, so be sure you quickly contact everyone you need to.
(j) Before April of each year, make sure any leaks, holes, or other damage to the embankments in your region has been fixed and that you have the necessary supplies and tools for any emergency repairs stashed in an easily accessible

location. The board will decide how much money to set aside and who will be in charge of carrying out the projects.

Alert and Warning Stage

(a) Sluice and lock gates should have guards to prevent theft.
(b) Send updates to higher-ups on the status of sluice gates, embankments, and installations when work is being done to fix them.
(c) Put safety measures in place to protect people, property, vehicles, and infrastructure.

Disaster Stage

(a) Maintain a permanent, round-the-clock (24/7) Flood Information Centre and assign a Liaison Officer to the regional Emergency Operations Center.
(b) Notify the local Disaster Control Room and the BWDB's Flood Control Cell if anything goes wrong.
(c) Help and aid the local civil government in their efforts to save lives, evacuate residents, and provide aid.
(d) The damaged infrastructure and supply lines need to be repaired, and this may be done via the careful management of technical personnel and the necessary supplies.
(e) As soon as the flood waters subside, you should take stock of the damage and draw up plans and designs for urgent repairs, rebuilding, and reinstallation that take both the near and long term into account. Funding and responsibility distribution will be handled by the BWDB.
(f) Work should be supervised by making frequent trips to the affected locations, with emergency preventative measures taken as needed.
(g) Seek the assistance of the IMDMCC through the local Civil Administration or the MoFDM in the event of a tough issue or scenario beyond the control of the respective office.
(h) In an emergency, you should do everything has to be done to save lives and property and to get people out of the area.

Rehabilitation Stage

(a) Determine the extent of the damage and draw up repair, reinstallation, or rebuilding plans as needed.
(b) Physical infrastructure, including sluice gates and water drains, should be restored and re-established as soon as possible with the assistance of local agencies and non-governmental organizations.
(c) Help the government and other organizations, such as NGOs, in their recovery efforts by working together.
(d) Create innovative project plans with the goal of averting flooding in certain places. Flood status reports will be sent daily from Water Development Board field offices to the following:

    (i) Control Room of the Ministry of Food and Disaster Management

(ii) Chairman, Water Development Board
(iii) Concerned Divisional Commissioner
(iv) All Deputy Commissioners of the affected areas
(v) All Upazila Nirbahi Officers of the affected areas

Ministry of Agriculture

The Secretary will be responsible for his or her personal work plan, as well as the standard obligations of the Ministry. Time must be used effectively for assessing catastrophe losses and making plans for agricultural restoration. The following responsibilities will be managed by the Secretary, Ministry of Agriculture.

Risk Reduction

(a) Appoint a high-ranking official as the Ministry's Disaster Management Point of Contact.
(b) Make sure that representatives from the NDMC, MoFDM, IMDMCC, District, Upazila, and Union DMCs all attend the appropriate meetings.
(c) Make sure all of its divisions participate in a sector-wide risk assessment and create a strategy to reduce risks.
(d) Make sure there is money set out in the budget to help carry out the action plan for lowering risks in that particular industry.
(e) Create an investigation plan to study the results of risk assessment in farming.
(f) Increase funding for agricultural research in order to develop new disaster-proof technologies.
(g) Instruct the relevant Ministry divisions to think about and include Disaster Risk Reduction Activities as they formulate their respective development strategies.
(h) Methodologies for incorporating a Disaster Management Plan into the development planning procedures of local-level offices should be created, approved, and implemented.
(i) Establish cooperative relationships with many organizations, particularly those devoted to research.
(j) Maintain a system for regularly revising Agriculture Extension's guidelines and manuals for use in the field.
(k) Help set up permanent training programs on disaster preparedness for the agricultural industry.
(l) Provide direction to DAE to raise public understanding of the value of disaster preparedness in agricultural settings.
(m) Make that the Ministry and its departments include disaster risk reduction in their plans, programs, and policies.
(n) Ensure that all relevant agencies are working together to reduce catastrophe risk.
(o) Increase interaction between national and international research institutions.
(p) Budget for risk mitigation initiatives such training and education, infrastructure maintenance and improvement, seed storage, and necessary analyses, etc.
(q) Set up and run a command center to monitor agricultural damage.

(r) When a cyclone or flood warning is issued, it is time to set up a command center and notify the appropriate parties so that seeds, fertilizer, pesticide, and other agricultural tools may be stored safely.
(s) Develop a Ministry-wide backup strategy for handling disasters and mitigating risks.

Emergency Response

Normal Times

(a) Build a contingency plan for the agriculture industry and keep it up to date.
(b) Preparedness plans should be evaluated every three months in accordance with the individual's work schedule.
(c) Set up efficient processes for receiving reports and sending out orders.
(d) Locate vulnerable regions that might be hit by the calamity.
(e) Get the impacted folks set up to receive their loan/grant seed, seedlings, fertilizer, and agri-implements.
(f) When moving stored seeds and tools, take the essential precautions.
(g) Prepare storage areas for planting materials including seeds, fertilizer, and pesticides.
(h) Get enough seed stock ready-to-plant in the affected regions.
(i) Prepare seasonal seed-starting beds and maintain a supply (to be done in places nearer to disaster areas).
(j) Make plans to store power tillers at the Upazila Headquarters in affected regions so that they may be used in times of necessity.
(k) Prepare different ranks of officers to take part in cyclone preparation at different stages.
(l) During the restoration phase, make plans to store enough supplies of all necessary goods in easily accessible locations.

Disaster Stage

(a) One Liaison Officer should be appointed to maintain communication with the MoFDM's EOC.
(b) Instruct lower-level offices to assist the Divisional Commissioner, Deputy Commissioners, the United Nations Observer, and the Chairman of the Union Parishad as needed.
(c) Plan a survey to assess the damage to your farm's crops and seed storage facilities so you can get started on a repair or rebuilding plan right away.

Rehabilitation Stage

(a) Determine the extent of the damage and loss in accordance with current regulations as soon as feasible, and complete the preparations for agricultural rehabilitation.
(b) Emergency distribution of agricultural materials and tools to the impacted region (s).
(c) Facilitate the plan's acceptance and funding distribution.

(d) Seeds, seedlings, fertilizer, and agri-implements, such as irrigation inputs, must be readily available and transported to the impacted populations in order to begin agricultural rehabilitation.
(e) You may use Bangladesh Bank to organize the delivery of agricultural supplies and agricultural loans.
(f) Gather all necessary resources for disaster management and rehabilitation programs to help those who have been impacted get back on their feet.
(g) All loans and help given to farmers must be used in a responsible manner.
(h) Provide NDMC and IMDMCC with a weekly update. As soon as the restoration plans are finished, a full report must be presented.
(i) Coordinate and distribute aid received from non-governmental organizations (NGOs) worldwide and inside the country, reporting all expenditures to the Ministry of Finance and Development Planning.
(j) Make sure that the relief distribution master rolls are prepared, kept, and submitted on time.
(k) Prepare a budget for the distribution of agricultural supplies like seeds and fertilizer after a natural catastrophe.

Department of Agricultural Extension (DAE)

The DG, Department of Agricultural Extension (DAE) will perform the following duties in addition to his normal functions.

Risk Reduction

(a) Create a Center of Excellence for Disaster Risk Reduction.
(b) Help the Agriculture Ministry create procedures for incorporating the Disaster Management Plan into the development planning processes of regional offices.
(c) Build alliances with relevant groups, such as academic institutions, nonprofits, and community-based organizations.
(d) The Agriculture Extension field manuals and guidelines are being brought up to date.
(e) Help set up permanent training programs on disaster preparedness for the agricultural industry.
(f) Create and execute measures to heighten agriculturalists' awareness of how to lessen their exposure to the hazards of natural disasters.
(g) Manage seed and seedling supplies, fertilizer and pesticide reserves, and other agricultural inputs and equipment to minimize potential disasters.
(h) Coordinate efforts to raise understanding and knowledge of disaster preparedness, hazard identification, and risk assessment within the agriculture sector and between personnel.

Emergency Response

Normal Times

(a) Appoint a DAE-based individual as the designated focal point for disaster management.

(b) Maintain a schedule of quarterly reviews of readiness under the guidelines laid forth in the Contingency Action Plan.
(c) Set up a Command Post and advise all relevant parties to secure agricultural supplies (seeds, fertilizer, pesticide, agri-implements, etc.) upon receipt of a cyclone/flood prediction notice.
(d) It is important to work with local authorities to locate high ground suitable for seedling beds in order to make up for potential losses in vulnerable regions.
(e) Do everything you can to prevent crop failure.
(f) Together with other involved offices/agencies, determine the probable impacted locations.
(g) Put in place the measures required to ensure the impacted individuals get seedlings/seeds, fertilizers, agri-implements, etc.
(h) Have a look at what you have on hand for things like seed, fertilizer, pesticides, and irrigation equipment.

Disaster Stage

(a) Manage your own Control Room and get a report on the state of your crops.
(b) Please notify the Ministry of Agriculture of any agricultural damage.
(c) Find out how much of your crops, livestock, tools, etc., were lost or damaged.
(d) Set up field offices at the Divisional, District, and Upazila levels to implement rehabilitation plans.
(e) Coordinate the timely delivery of seed, seedlings, fertilizer, pesticide, agri-implements, and other agricultural supplies to impacted regions.
(f) You should tell the DC, the UNO, and the Union Parishad Chairman to provide their full support to everyone who needs it.
(g) Farmers' access to agricultural loans will be determined and guaranteed by the Upazila Agri Loan Committee.
(h) A farm's inputs and tools should be dispersed.

Rehabilitation Stage

(a) Ascertain the amount of money and resources needed to restore the damaged regions.
(b) Make a concerted effort via local government to provide seed, seedlings, fertilizer, agri-implements, etc., to the afflicted regions immediately.
(c) Quickly organize the delivery of agricultural supplies and equipment.
(d) Plan agri-rehabilitation assistance for farmers via local politicians.
(e) Inform the public about agri-rehabilitation using media like radio, TV, and newspapers.
(f) Through education, inspection, and oversight, local field authorities may make sure aid, assistance, and loans are used fairly.
(g) The Ministry of Agriculture should get a final report once weekly updates on restoration work have been sent.
(h) After determining the need for supplementary agricultural inputs, the Ministry of Agriculture should reach out to foreign and non-profit organizations for help.

Field Offices of the DAE

The field offices of the DAE at Divisional, District, and Upazila levels will perform the following duties in addition to their own duties.

Risk Reduction

(a) You should try to make it to as many UzDMC meetings as possible.
(b) Assess the potential dangers facing the region's farmers, then draw up a strategy for mitigating those risks.
(c) Reduce existing and future threats to the agriculture industry by ensuring their successful execution.
(d) Build alliances with relevant groups, such as academic institutions, nonprofits, and community-based organizations.
(e) Prepare for future hazards, such as those posed by climate change, by organizing demonstrations in the field.
(f) Manage seed and seedling supplies, fertilizer and pesticide reserves, and other agricultural inputs and equipment to minimize potential disasters.
(g) Plan events to educate farmers and other agricultural workers on hazard identification and risk assessment.

Emergency Response

Normal Times

(a) Conduct control room operations and other duties as assigned by the DG, DAE.
(b) In order to grow enough seedlings to fulfill demand in impacted regions, it is necessary to first locate and then choose suitable high ground within the region.
(c) Act sensibly in the face of catastrophic crop failure in the region.
(d) Together with other departments, agencies, and organizations, you must determine which regions will be hit worst by the catastrophe.
(e) Plan ahead to ensure that individuals in the impacted region have access to agricultural necessities including seeds, seedlings, fertilizer, and tools.
(f) After assessing the available supply, you should submit a request to DAE for the immediate purchase of seed, fertilizer, pesticides, and other agri-implements/inputs.

Disaster Stage

(a) Perform duties in the Control Room to gather data on crop failure.
(b) Report any crop failures or damage to seedlings to the Agriculture Ministry.
(c) Calculate the total amount of damage or loss to crops, cattle, and machinery.
(d) Initiate restoration plans through district and Upazila-level field offices.
(e) Seeds, seedlings, fertilizer, pesticide, and agricultural equipment should be moved to the most centrally placed storage facilities in the affected regions to ensure rapid distribution.
(f) Lend a hand to the local government in their efforts to evacuate the area and save lives.

(g) It is your responsibility to make sure the Upazila agri-loan committee knows they will be receiving their crop loan and that they are really needed.
(h) A farm's inputs and tools should be dispersed.

Rehabilitation Stage

(a) After calculating the extent of the damage, you should submit a plan for the financial and material restoration of the impacted regions to the relevant authorities.
(b) Coordinate the distribution of agricultural resources such as seeds, seedlings, fertilizers, and agri-implements to impacted regions as quickly as possible so that agricultural production may be restored there.
(c) Rapid distribution of agricultural supplies and tools is essential for farmers' recovery.
(d) Farmers should be trained and visited in the field to learn about rehabilitation issues.

Agricultural Information Services (AIS)

(a) Information on drought, flood, flash flood, and salinity will be collected from various early warning centers, disseminated to the community level, and copied to DMB so that appropriate steps may be taken to mitigate their effects or adapt to them.

Bangladesh Agricultural Development Corporation (BADC)

The Chairman, BADC will ensure the following in addition to his normal duties.

Risk Reduction

(a) Use farmers' field demos to learn about cutting-edge technology that can adapt to changing weather patterns.
(b) In order to test and evaluate new technical possibilities, it is essential to forge close ties with all agricultural research organizations and extension departments.
(c) Build and keep updated a database of organizations doing agricultural research and extension.
(d) Seed inventory should be organized to meet seasonal demands.
(e) The timely and affordable delivery of agricultural supplies to farmers is a priority.

Emergency Response

Normal Times

(a) Choose a person to serve as the focal point at the Corporation's headquarters.
(b) Set up disaster management programs with local points of contact.
(c) Monitor the readiness of your subsidiary and affiliated offices once every three months using your own Action Plan.
(d) Make sure that everyone who has to be is prepared for the recent severe crop catastrophe.

(e) Get the impacted folks set up with a loan or grant so they can get seeds and seedlings.
(f) Get ready to relocate your seed supply to a more secure location.
(g) Seed inventory should be organized to meet seasonal demands.
(h) Plan for the availability of repair components for both deep and shallow tube wells.

Disaster Stage

(a) Keep in regular contact with the Ministry of Agriculture and run the Control Room.
(b) Collect data on damages and losses, and report them to the Agriculture Ministry.
(c) Create a plan for rehabilitation, evaluate it, and find the money to pay for it.
(d) Make plans for the storage and rapid distribution of seed, seedlings, and other plant materials by transferring and transporting them to a central location.
(e) Lend a hand to the local government in their efforts to evacuate the area and save lives.

Rehabilitation Stage

(a) After a quick evaluation of the loss or damage to agricultural works dependent on irrigation, prepare plans for repair.
(b) Efforts should be made to quickly provide goods like seeds and seedlings to those in need in order to begin the process of rehabilitation in the impacted regions.
(c) When it comes to fixing the deep tube wells and water pumps used for irrigation, please lend a hand to the farmers in any manner you can.

Ministry of Fisheries and Livestock

The following are additional obligations that fall within the purview of the Secretary of Ministry. The Ministry must also implement effective strategies for preventing and controlling Avian Influenza and other emerging zoonotic illnesses that pose a threat to human health.

Risk Reduction

(a) Make a decision on who will serve as the Ministry's Disaster Focal Point.
(b) Assess the sector-level risk of bird flu and other zoonotic illnesses, and devise a strategy for mitigating that risk.
(c) Make sure the Ministry has enough money in its budget to carry out the strategy.
(d) Integrate disaster management within MoFL's national development strategy.
(e) Plan a study on the effects of risk assessment on the fishing and cattle industries.
(f) Be sure the Ministry has enough money in its budget to deal with disasters.
(g) Put into action the policy of conducting risk assessment and risk reduction actions that influence Ministry services.
(h) Specifically address Avian Flu in your efforts to prevent and manage emerging Zoonotic illnesses that pose a threat to human health.

(i) Involve the cattle and fishing businesses in risk assessment and risk reduction efforts via training, awareness, and education initiatives.
(j) Oversee emergency planning tasks include finding high ground where animals might take refuge and storing away supplies of essentials like vaccinations and medications for livestock and poultry.
(k) In high-risk locations, it is important to prepare and monitor an inventory of livestock and poultry.
(l) Create plans and structures for handling livestock in times of crisis, such as emergency feeding, evacuating animals, and rehabilitating the industry after a disaster.
(m) Safety in the trawler fleet may be improved by mandating registration with the Marine Fisheries Department and the installation of radios, wireless communication devices, and life vests on all vessels.
(n) Take action to lower the danger of drinking saltwater by working with the BWDB to improve the height and strength of government embankments and sluice gates.
(o) Power-driven pumps for draining coastal ponds should be made readily available in conjunction with the Bangladesh Agricultural Development Board.
(p) As the cyclone and flood seasons approach, it is important to educate fisherman about the potential dangers they face and the steps they may take to mitigate those risks.
(q) Create a Ministry of Defense readiness strategy plan for mitigating risks in certain industries.
(r) Make a strategy for the Ministry's reaction to emergencies in certain sectors.
(s) Put in place a Ministry of Risk Management and Communication risk communication system.
(t) Develop a Ministry-wide backup strategy for handling disasters and mitigating risks.
(u) Make that the Ministry and its departments include disaster risk reduction in their plans, programs, and policies.
(v) Keep all of the government agencies involved in reducing catastrophe risk in sync with one another.

Emergency Response

Normal Times

(a) Security and safety of precious assets may be ensured by making good use of time throughout the rehabilitation period, which should be reviewed every three months according to one's own Contingency Action Plan.
(b) Make the necessary preparations for the receipt and distribution of reports.
(c) Find the likely places that will be hit by the calamity.
(d) Start a contingency fund solely for post-disaster recovery expenses.
(e) Fisheries and livestock authorities should get training in disaster preparation, environmental loss and damage, and restoration.

Alert and Warning Stage

(a) Direct subordinates at all levels of the Directorate under his command to implement and maintain stringent safety protocols for the Directorate's properties and assets.
(b) One Liaison Officer should be appointed to maintain communication with the Emergency Operations Center (EOC) of the DM&RD and the Disaster Management Bureau.
(c) You should tell everyone in the Directorate to help the Divisional Commissioners and Deputy Commissioners, the Chairman of the Upazila Disaster Management Coordination Committee, the Chairman of the UDMC, and the CPP Officials in any way they can.

Rehabilitation Stage

(a) Inspect and survey all lost or damaged property, including livestock, poultry, fish, fish hatcheries, fish ponds, trawlers, and other structures, as soon as possible.
(b) Complete the assessment of losses and damage to cattle, poultry, fish farms, hatcheries, fishing trawlers, training and research institutions, medicines and chemicals, and submit a report to the relevant authorities in order to get necessary funding.
(c) Pay for the plans that have been authorized.
(d) Reconstruction efforts should be coordinated with local authorities and relevant government departments.
(e) Get up a report on the finished relief and rehabilitation programs and send it into the National Disaster Management Council.
(f) Prepare for the possible need to import cattle in the course of restoration efforts.
(g) Protect, relieve, and rehabilitate livestock or fisheries stock, as necessary, by assisting impacted people, fishermen, pisciculturists, and farmers with health care, feeding, stocking, and protection activities on the ground.
(h) Aid the relief and reconstruction efforts in the cattle and fishing sectors by providing expert assistance on these matters.

Department of Livestock Services (DLS)

The department of Livestock Services (DLS) will perform the following duties in addition to normal duties.

Risk Reduction

(a) Choose a Point of Contact for Emergency Preparedness.
(b) The DLS strategy should take into account the risks posed by disasters and climate change.
(c) Put the plan into motion.
(d) Conduct risk assessment and risk reduction actions, such as bio-security management on livestock and poultry farms, along with the poultry and livestock industries via training, awareness, and education initiatives.

(e) Oversee emergency planning tasks include finding high ground where animals might take refuge and storing away supplies of essentials like vaccinations and medications for livestock and poultry.
(f) In high-risk locations, it is important to prepare and monitor an inventory of livestock and poultry.
(g) Create plans and structures for handling livestock in times of crisis, such as emergency feeding, evacuating animals, and rehabilitating the industry after a disaster.
(h) Provide assistance to the Ministry of Fisheries and Livestock and help them develop a sectoral contingency plan for their disaster management and risk reduction efforts.

Emergency Response

Normal Times

(a) Please appoint one Directorate Liaison Officer to serve as the department's primary point of contact for disaster management.
(b) Keep in touch with subordinate offices, officials from pertinent field-level government offices, and CPP, and conduct a review of preparation every three months based on the Directorate's own Contingency Action Plan.
(c) Before the start of the cyclone/flood season, notify all field officers in the affected cyclone/flood-prone region to take precautions for the protection of own assets, such as stocks in poultry farms and shelter place for animals in the cattle farms.
(d) To safeguard cattle from devastating floods during the flood season and tidal bores during cyclone season, it is necessary to choose and designate high areas in coordination with local government and CPP.
(e) Prepare emergency supplies of medications and other items to combat and contain infectious illnesses that might threaten livestock and poultry.
(f) Get a count of the animals and birds and make an inventory in places prone to cyclones and floods. Repeatedly refresh the same.
(g) Assist the local government in moving cattle to safer areas and making arrangements for vaccination and treatment in the event of an impending hurricane or tidal bore.
(h) In the event of livestock loss, make plans for their rehabilitation and compensation.
(i) Start plans to provide emergency animal feed to the impacted regions as soon as possible.
(j) Prepare local and intermediate-level authorities and workers for storms, floods, pollution, and construction by providing them with training in these areas.

Disaster Stage

(a) Establish communication between the DLS and the MoFDM's EOC via the assignment of a single Liaison Officer.
(b) Help the local government with activities during floods that include moving or rescuing animals.

(c) Prepare safe havens for animals and poultry by having them vaccinated and treated quickly in the event of floods.
(d) Get the animals in the affected regions vaccinated (before floods).

Rehabilitation Stage

(a) Coordinate with other relevant Ministries/Departments and be ready to take immediate action on the loaned purchase of cattle and the delivery of animal fodder as emergency aid.
(b) Quickly conduct a survey to determine the extent of the damage, and if required, bring in replacement cattle.
(c) Assemble emergency veterinary response teams and send them to the impacted region.
(d) Make contingency plans to replace the wealth lost in cattle, and prepare for the livestock's recuperation. The DLS has set aside an indefinite treasury for this exact function.
(e) Plan how to quickly get animal feed for distribution in impacted regions. In case of an unexpected expense, the Directorate of Livestock will make the necessary arrangements.
(f) Report any missing or dead livestock or poultry, as well as any cases of sickness in livestock or poultry, to the International Veterinary, Disease, and Animal Care Coordination Center (IMDMCC).
(g) In the event of flooding, please inform IMDMCC of the locations where animals have been moved or saved.
(h) Help out right away by coordinating with local authorities to set up a system of emergency aid and recovery.
(i) Help the local government get the cattle back to their homes and to the owners who evacuated with them.
(j) Help the affected folks stock up on feed for their animals and birds until regular supplies are restored.
(k) Put into action all plans immediately to help rehabilitate cattle, including the animals that were previously chosen.
(l) Livestock and poultry import strategies should be put into action and distributed to impacted communities.

Field Offices of the DLS

In addition to their own responsibilities the offices of DLS at District, Upazila, and Union level will perform the following duties within their respective areas.

Risk Reduction

(a) Make sure you have a designated Disaster Management Focal Point who attends and takes notes at all Disaster Management Committee Meetings.
(b) Create a strategy to mitigate the threats to the cattle industry on a regional level based on what you learn.

(c) Use the funds allotted in the Annual Development Programme to carry out the plan of action.
(d) In high-risk locations, it is important to prepare and monitor an inventory of livestock and poultry.
(e) Conduct risk assessment and risk reduction training, awareness, and education programs for small and medium poultry and animal farms.
(f) Develop a backup strategy for emergency situations including local disaster management and risk reduction initiatives.

Emergency Response

Normal Times

(a) Preparations for the protection of livestock, poultry, and domestic animals and their feed are reviewed annually in April by all field-level officials of the Directorate and the farmers, just in time for the onset of cyclone season.
(b) Examine the readiness measures according to the Contingency Action Plan in the cyclone/tidal bore-prone regions with the subordinate offices, CPP, and farmers' representatives to ensure the animal wealth is kept under a disciplined safety arrangement beginning at the lowest level.
(c) In the event of a cyclone or tidal bore, it is important to pick high ground, hillocks, or reinforced earthen mounds in collaboration with the local authority and to make this information well known in the area.
(d) Plan for the unexpected by storing medications and tools for treating and preventing infectious illnesses in livestock and poultry in regions at risk from cyclones and tidal bores.
(e) In tidal bore and cyclone-prone locations, it's important to have a supply of emergency animal feed on hand.
(f) Livestock and poultry censuses should be conducted annually in April in cyclone and tidal bore-prone regions.
(g) Set up orientation sessions for local authorities and employees to get them ready for a hurricane.

Disaster Stage

(a) One person should be appointed as the Disaster Control Room's Liaison Officer.
(b) Prepare rescue and evacuation activities for trapped livestock and poultry by coordinating with local government and people/agencies like Union Parishad.
(c) Vaccinate and treat animals and poultry in safe locations before the floods hit.
(d) Animals in flood-prone regions should be vaccinated in masse.

Rehabilitation Stage

(a) Make plans for a loan or grant to help people in vulnerable regions buy animals and stock up on feed.
(b) As soon as possible, you should conduct a survey to assess the level of damage, losses, and requirements, and you should also organize for the import of animals from other places.

(c) Animals in the afflicted regions should be treated immediately by field teams.
(d) Please provide the department head with a comprehensive report detailing the total number of animals that were either lost or became ill.
(e) Find out where all the livestock shelters are and how many animals have been moved or gathered recently.
(f) Help and collaborate with local authorities in planning emergency relief and restoration efforts.
(g) Make contingency plans on a local level to replace lost animals and poultry and restore those that have been harmed.
(h) Animal feed and other feed should be stockpiled in preparation for emergency distribution in the impacted regions.
(i) Help farmers get their animals back from temporary shelters so they may be reunited with their families.
(j) Help the affected individuals get cattle and poultry feed.
(k) Make plans to acquire livestock and poultry, including chosen animals, using loans from the Bangladesh Bank in order to aid in their recovery.

Directorate of Fisheries (DoF)

The Directorate of Fisheries will perform the following duties in addition to its normal functions.

Risk reduction

(a) Pick one person to serve as the Disaster Management Focal Point and let them know they have the support of the DMB.
(b) The Fisheries Directorate's plan of action should take into account the possibility of natural disasters.
(c) Put the plan into motion.
(d) Fish producers should participate in training, awareness, and education programs focusing on risk assessment and risk reduction.
(e) Provide assistance to the Ministry of Fisheries and Livestock and help them develop a sectoral contingency plan for their disaster management and risk reduction efforts.

Emergency Response

Normal Times

(a) Protect pisciculture ponds, chemicals, implements, water and road transportation, and own assets in various DoF offices by having all involved field officers take safety precautions long before the start of flood season.
(b) You should make sure that each fishing trawler has a wireless and radio set, and that the fishing vessels are registered with the Marine Fisheries Department, before you give them a fishing license.
(c) As such, appropriate administrative and legal measures should be taken to guarantee that all fishing boats and trawlers operating in the Bay of Bengal are equipped with a radio receiving set and enough life jackets for all crew members.

(d) Generate and maintain an up-to-date inventory of public and private fisheries assets in cyclone/tidal bore vulnerable locations.
(e) Compile a proper Upazila-by-Upazila census report and update it at regular intervals.
(f) Make and keep track of a list of all fishing boats that are seaworthy (including the owners' contact information).
(g) Government embankments and sluice gates need to be built at the right height and with enough strength to resist the entrance of saltwater caused by tidal bore. This should be done in conjunction with the Bangladesh Water Development Board.
(h) In order to remove saltwater from coastal ponds, it will be necessary to coordinate with BADC to find power-powered pumps.
(i) Prepare the local cops, personnel, and fishermen community to respond to and recover from cyclones by providing them with the necessary training and orientation. Assemble training for MFL and DoF field officers and workers.

Disaster Stage

(a) Take the time to consult with BWDB to ensure that the embankment sluice gates are robust enough to resist the pressure of waves caused by tidal bore and to prevent the intrusion of saltwater.
(b) Start counting the fish that have been lost right now, and go to work on some long-term relief and rehabilitation plans for the commercial and recreational fishing industries.
(c) Help the local government secure ocean-going boats for search-and-rescue missions.
(d) Maintain the Directorate's Control Room and dispatch a Liaison Officer to coordinate with the area's DMC.

Rehabilitation Stage

(a) The availability of power pumps to remove saltwater from public and private ponds should be coordinated with the local government and the Bangladesh Agricultural Development Corporation (In case of cyclone disaster).
(b) It is essential that a plan for the long-term alleviation and rehabilitation of publicly and privately held fishing resources be formulated and put into action.
(c) Make an inventory of the impacted fish farmers and fishermen.
(d) Help private aquaculturists recover their ponds and farms by giving them access to expert technical help.
(e) Help and collaborate with local authorities to implement relief and recovery initiatives.
(f) Establish schemes to provide financial aid to fishermen and aquaculture workers who have been impacted.
(g) Help pond owners who are struggling with fish supply and pisciculture by offering them expert advice.

Field Offices of Fisheries Directorate

The following responsibilities will be taken on by the field officers of the Fisheries Directorate at the Divisional, District, and Upazila levels and of projects.

Risk Reduction

(a) Appoint a representative to serve as the "Focal Point" for Disaster Management at all DDMC Conference Calls.
(b) Create a strategy to mitigate the threats to the cattle industry on a regional level based on what you learn.
(c) Use the funds allotted in the Annual Development Programme to carry out the plan of action.
(d) In high-risk locations, it is important to prepare and monitor an inventory of livestock and poultry.
(e) Conduct risk assessment and risk reduction training, awareness, and education programs for small and medium poultry and animal farms.
(f) Develop a backup strategy for emergency situations including local disaster management and risk reduction initiatives.

Emergency Response

Normal Times

(a) Farmers and fishermen in cyclone- and flood-prone regions will get annual warnings from DoF field officers about how to prepare their boats, trawlers, fishing gear, fishing equipment, fish fry, and stock hatchery for the next season.
(b) Meet with subordinate offices, CPP Pisciculturists, and fishermen representatives to assess readiness in light of the DoF's own Contingency Action Plan.
(c) Find a secure location to store fishing vessels, trawlers, and equipment for the duration of the tragedy.
(d) Before issuing a fishing license, check each trawler for the presence of a wireless and radio set.
(e) As part of the administrative process, make sure that all Bay of Bengal fishing vessels and trawlers have working wireless receivers and enough life jackets for everyone on board.
(f) Make and keep up-to-date a list of all government and private fisheries assets in disaster zones.
(g) Keep data on the number of fishermen, pisciculturists, and fish farms in regions hit by catastrophe.
(h) Keep track of the fishing boats, trawlers, and ocean-going vessels in the region, as well as their owners and captains.
(i) Keep in constant contact with BWDB employees to guarantee the proper functioning of embankments and sluice gates in affected regions, preventing saltwater from seeping in under the influence of tidal bore pressure.
(j) See to it that the sluice gates are working properly.

(k) Working with CPP, coordinate with local BADC authorities to provide access to power pumps for draining saltwater from ponds.
(l) Learn about storm evacuation plans, damage assessments, and restoration techniques.

Disaster Stage

(a) Help the local government acquire ocean-going boats for use in relief, rehabilitation, and rescue operations in the aftermath of a storm.
(b) To contact the local DMC, please activate the Control Room and dispatch the Liaison Officer.

Rehabilitation Stage

(a) As soon as possible, submit the competent authorities the results of a survey to determine the extent of the loss or damage to public and privately held fishing assets, as well as any plans you've made for long-term relief and rehabilitation.
(b) Work with the regional government and BADC to arrange the import of the necessary quantity of power pumps using public and private funding, if at all feasible.
(c) Keep higher-ups updated on the state of the fisheries development sector's long-term relief and rehabilitation efforts.
(d) Motivate and help the pond fish growers who were impacted to start over again.
(e) Help the government in the area with disaster relief and restoration efforts.
(f) Assist fishermen and aquaculturists in securing a loan.

Ministry of Health and Family Welfare

Curative and preventative health services shall be provided to the public by the officials and personnel of the Ministry of Health and Family Welfare, Directorate of Health Services, and its connected organizations. The Ministry will also keep a close watch on the rapid deployment of medical personnel to disaster zones in order to restore health services as quickly as possible following an incident. For the Health Directorate to fulfill its mandate from the Ministry of Health and Family Welfare, it must carry out the following responsibilities.

Risk Reduction

(a) Appoint a high-ranking official in this Ministry to serve as the main contact.
(b) You must attend every NDMC, IMDMCC, and MoFDM meetings.
(c) Instruct all relevant sections to implement the decisions made at NDMC, IMDMCC, and MoFDM meetings that pertain to their duties.
(d) Ensure that the Ministry of Health and Family Welfare's plans, programs, and initiatives take disaster risk reduction into account.
(e) Make that the Ministry's risk management and mitigation efforts are adequately funded.
(f) Create a Ministry-wide Risk Reduction Communication System.
(g) Establishing an efficient monitoring and evaluation system for risk reduction activities in the health sector.

(h) Create a backup strategy for the Ministry's risk mitigation and management initiatives.
(i) Coordination of emergency planning efforts in high-risk regions with the Geologist and MoFDM.
(j) Volunteers, members of the community defense group, Ansar, primary school teachers, and religious leaders should all get training in first aid and life safety procedures as part of the earthquake preparation program.
(k) Create a strategy for setting up makeshift hospitals in the event that permanent ones are destroyed by an earthquake.
(l) Backup facilities for earthquakes and other calamities, as well as modifying the hospital's infrastructure to withstand an earthquake, will protect the hospital's lifelines and infrastructure.

Emergency Response

Normal Times

(a) If you live in an area prone to earthquakes, it's important to have a plan in place and to update it annually.
(b) Coordinate with relevant offices/departments and conduct an annual April and September review of the Ministry's cyclone preparation Contingency Action Plan. Participate in disaster area mobilization exercises with the CPP and other relevant organizations.
(c) Cyclone preparation volunteers, village defense party members, Ansars, primary school teachers, and religious leaders should all get training in the use of oral saline, first aid, and illness prevention.
(d) Make sure that the Upazila Health Centres in affected regions have sufficient supplies of ambulances, medication, vaccinations, surgical equipment, etc.
(e) Make an Upazila-by-Upazila list of all the medical and paramedical staff in the affected region and submit it to the relevant DCs.
(f) Make sure the Upazila Health Complex is ready to handle any extra patient care needs that may arise during and after a catastrophe.
(g) During and after natural disasters like floods and cyclones, it is important to provide healthcare education via the media.
(h) In disaster zones, organize a medical team with the necessary supplies.
(i) In order to satisfy the urgent medical care requirements, temporary hospitals should be set up in Health Sub-Centres and cyclone shelters. If cyclones wipe out conventional medical facilities, people should put up makeshift hospitals in bomb shelters.
(j) Participate actively in IMDMCC meetings and collaborate with other departments' health-related initiatives.
(k) Appoint one Ministry employee as the department's primary contact for all matters related to disaster management.
(l) Create a strategy for managing disasters in this Ministry and its many departments.

(m) Create a Ministry-wide emergency response strategy including all relevant sectors.

Alert and Warning Stage

(a) It is the responsibility of the Director General of the Health Directorate to organize the rapid deployment of medical professionals, as well as the rapid delivery of necessary supplies to the region under attack.

Disaster Stage

(b) One Liaison Officer should be appointed to serve as the primary point of contact with the MoFDM's EOC.
(c) Establish a Command Center and maintain it with full-time staff (24 h).
(d) Verify that the DG Health Services is carrying out his responsibilities as outlined in paragraph 10(1) below.
(e) Prepare shelters for storm evacuees by providing them with food and lodging.
(f) Oversee the operation of primary healthcare centers for impacted communities.
(g) Take care of hospitalized people in regions hit by the disaster, whether it means transporting them to a safe location or another hospital.
(h) Determine the best ways to prevent contamination of drinking water sources and evaluate all potential sources.
(i) Oversee the distribution of clean water in cyclone shelters and other evacuation centers, as well as in impacted communities. This may include the distribution of water purification pills or bleaching powder.
(j) Oversee the medical and health departments involved in emergency response, disaster relief, and rebuilding efforts.

Rehabilitation Stage

(a) Avoid complacency in the face of a potential or actual epidemic outbreak and instead take swift and decisive action to stop it in its tracks.
(b) Get ahead of potential outbreaks of diseases like typhoid and cholera by taking prophylactic action.
(c) In order to keep the National Disaster Management Council apprised of health-related efforts in the impacted regions, please submit a report.
(d) Allocate and manage resources in accordance with established budgets.

Directorate General of Health Services (DGHS)

Directorate General of Health Services (DGHS) shall ensure performance of the following duties in addition to its normal functions.

Risk Reduction

(a) Help the Ministry of Health and Family Welfare make sectoral policies more disaster-resistant by factoring in the potential impact of earthquakes and other hazards.

(b) Perform Ministry-wide risk analysis, and develop and put into effect a strategy for mitigating risks in specific sectors. Protect the healthcare system's infrastructure and facilities against physical and potential threats by taking the necessary steps.
(c) Conduct risk analysis and risk reduction measures, such as an evaluation of the healthcare system's vulnerabilities.
(d) Conduct public health, mass casualty management (MCM), and related awareness, training, and education programs, focusing on topics including public health, hygiene, sanitation, and clean drinking water.
(e) Conversations with the geological survey crew to confirm delivery of the seismic danger maps.
(f) Make sure you have a plan in place for reliable medical treatment in the event of an earthquake.
(g) Implement a plan to get medical aid out to the afflicted population and into the hospitals where they live.
(h) Be sure that there is a reliable source of clean water and power near any emergency shelters you set up in the impacted regions, and that a mobile or field hospital is staffed with qualified medical personnel.
(i) Create a central database to help emergency medical services better assist patients.

Emergency Response

Normal Times

(a) If you live in an area prone to earthquakes, it's important to have a plan in place and to update it annually.
(b) Cyclone, flood, and earthquake preparation plans, as well as plans for a backup Medical Team, medication, vaccines, and other supplies, should be reviewed every three months with the officers and personnel of subordinate offices.
(c) One Directorate General of Health employee should be appointed as the point person.
(d) Construct a makeshift medical corps prepared to respond to emergencies with the necessary drugs, supplies, and equipment.
(e) Act in concert with the local government at all levels to implement the Directorate's priorities.
(f) Examine whether or not cyclone-prone communities have enough supplies of medication and surgical equipment.
(g) Gather information on available medical and paramedical personnel, and update the list annually.
(h) Prepare the cyclone shelters with food and lodging for the displaced.
(i) Arrange for the care of wounded people and make sure there is clean water, medication, and vaccines available in the cyclone shelter.
(j) Get the word out about health care options via the media (including TV, radio, and print).

(k) Find out how many more workers, supplies, and healthcare items will be needed, and set aside the necessary money from the budget.
(l) Water and electricity might suddenly go out during a crisis. So, it's important to make sure there are backup plans in place for the utilities.
(m) A central data repository will be created by the Health Directorate, and an adequate supply of essential medications will be made available in the event of a catastrophe.

Alert and Warning Stage

(a) As soon as a cyclone/flood danger warning is received from the MoFDM/Storm Warning Centre/FFWC, officers/staff of the DG Health Services of disaster zones shall be alerted.
(b) Conduct Control Room operations and keep the IMDMCC and the MoFDM apprised
(c) Coordinate with the DMB to deploy a medical response unit to the worst-hit regions.
(d) Coordinate with the appropriate local authorities to set up any extra road and water transports that are required, in addition to your own vehicles.
(e) Inform the regional offices that their supplies, medications, and property may be at risk.
(f) You should appoint a Liaison Officer whose job it is to maintain communication between the EOC of DMB and the Ministry of Health and Family Welfare.

Disaster Stage

(a) Take shifts in the Control Room (24 h).
(b) Provide impacted individuals with urgent primary medical treatment and, if necessary, dispatch a medical response team (s).
(c) Make plans to transport sick and wounded people to a safe location, such as a hospital or field hospital, and provide assistance once they arrive.
(d) The provision of water purification tablets, bleaching powder, etc., and the rigorous observance of health care and health service norms in cyclone shelters will guarantee the availability of clean drinking water.
(e) Verify that safeguards have been put in place to prevent pollution from reaching drinking water sources, and make any other required preparations.
(f) Increase the number of available beds and medical services at surrounding facilities.
(g) Do everything you can to avoid getting cholera and typhoid.
(h) Protect yourself and others against illness by continually being on the lookout for signs of an epidemic and acting quickly when one is detected.
(i) The DMB should receive a daily report on the number of casualties among the sick and wounded, with a copy sent to the Ministry of Health and Family Welfare.

Rehabilitation Stage

(a) Prepare for cholera and typhoid prevention measures to continue until the end of the vaccine campaign in the afflicted regions.
(b) Keep treating those who are currently being treated.
(c) Protect yourself and your community against the spread of any epidemic by being on high alert for their emergence.
(d) Maintain your focus on improving healthcare, educating the public, and limiting the population.
(e) As soon as saltwater recedes from the damaged regions, recleaning of the water supply should begin.
(f) Get in touch with the IMDMCC and give them a full account of what you did.
(g) Be sure to keep an eye on the nutrition levels of those impacted and help them out as much as you can.

Field Office of the DG Health Services

In times of crisis, the District Civil Surgeon and the Health Administrator of the Upazila Health Complex would be expected to go above and beyond the scope of their customary responsibilities.

Risk Reduction

(a) Upazila-level management of health care facility and service readiness, including ambulance, medication, vaccinations, and surgical equipment.
(b) Instruct the community members on the use of oral saline, basic first aid, and health promotion.
(c) Collect data on medical and paramedical workers in high-risk locations.
(d) Health Sub-Centres and emergency cyclone shelters will need medical care, and thus it is important to prepare for this in advance.
(e) Guarantee electrical and communication services in vulnerable locations, and create backup plans in case the primary ones fail.
(f) Building everyone's understanding of earthquake preparedness so that they can respond effectively.
(g) Create a comprehensive list of organizations that may participate in emergency management, including blood banks.
(h) Help the government in the area pinpoint potentially dangerous spots.
(i) Maintaining a steady supply of potable water requires coordination between the offices of public health and engineering. Help guide the NITOR's efforts to restore the mobility of those rendered immobile by the earthquake.

Emergency Response

Normal Times

(a) In the event of an earthquake, you should always be prepared.
(b) Keep complete and up-to-date information on all medical and paramedical staff (public and private) each year. Make plans for on-the-spot training to improve the effectiveness of medical personnel. Every three months, evaluate the status

of catastrophe supplies, pharmaceuticals, and personnel, and make adjustments as needed to make up the shortfall.

(c) In the event of a crisis, it is imperative that you double-check the stockpiles of usable medical equipment and pharmaceuticals.
(d) In disaster-prone places, it's important to take stock of medical resources including emergency response teams, medication, vaccinations, supplies, and equipment, and where they're stored.
(e) Prepare for the running of temporary and mobile hospitals, drawing on prior experience if required.
(f) Communicate with the authorities there.
(g) Protect against the risks of cyclones and floods by securing supplies, medications, paperwork, and transportation.

Alert and Warning Stage

(a) Notify all parties involved, activate the Control Room, and appoint a Liaison Officer to liaise with the local DMC(s).
(b) A number of medical teams should be on standby, ready to be sent to impacted regions or threatening areas should the need arise.
(c) Make plans to request, when needed, more resources and personnel from the appropriate superior.
(d) Make the required transportation arrangements through land and sea using departmental resources and the help of the local government.
(e) Cyclone shelters, health subcenters, and relief centers should all be inspected for cleanliness and adequate medical treatment.
(f) Make sure that sick people, the vulnerable, and the troubled all have access to the same food and lodging.
(g) Prepare rural communities for emergencies by having departmental staff, NGOs, and local government institutions teach residents about health, sanitation, and the environment.
(h) Protect medical facilities and supplies by moving them to a cyclone shelter if required.

Disaster Stage

(a) Maintain the Control Room's operations around-the-clock (24/7) and provide relevant data to the officer in charge.
(b) One Liaison Officer should be sent to the local DM Command Center.
(c) Make plans to run the Health Complex and Health Sub-Centre as required by recovering the supplementary machinery, personnel, and useable materials via enough supply for the urgent maintenance.
(d) It is vital that impacted individuals have access to routine primary care. Make arrangements for extra beds and medical attention at neighboring hospitals if needed.
(e) People who are sick or wounded should be moved to storm shelters or hospitals or Health Sub-Centres.

(f) Water purification pills, bleaching powder, etc., should also be supplied with potable water to guarantee that temporary hospitals, cyclone shelters, and health Sub-Centres are kept sanitary.
(g) To guarantee that drinking water is not polluted, it is important to investigate all of its potential sources.
(h) Prepare to vaccinate the selected population.
(i) To stop the spread of an epidemic, start a public relations campaign.
(j) Keep the higher-ups updated with a daily tally of the casualties suffered from sickness and other diseases.
(k) Gather and save the necessary paperwork on the death, then forward it on to the right people.

Rehabilitation Stage

(a) Stay vigilant and take precautions to stop the spread of any epidemics.
(b) Keep up the good work of spreading useful information on health and family planning via the efforts of your department's staff and non-governmental organizations.
(c) Remove any contaminants from the area's drinking water supplies.
(d) Make sure the proper people get the whole report.
(e) Act on initiatives to improve access to health care and reduce hunger.

Ministry of Environment and Forest

Environmental damage may occur as a result of natural catastrophes. Degradation of the natural world may potentially lead to catastrophic events. Disasters like cyclones, floods, and droughts may be mitigated through reforesting. Therefore, the Ministry of Environment and Forests plays an important role in both the normal period and the rehabilitation phase of disaster management and risk reduction owing to climate change. The following tasks will be carried out by the Ministry of Environment and Forests through two departments reporting to it, such as the Forest and Environment Department.

Risk Reduction

(a) Appoint a senior Ministry employee to serve as the Focal Point and facilitate communication between the Ministry and the National Disaster Management Committee (NDMC), the Missouri Disaster Management Committee (MOFDM), and the International Disaster Management Coordinating Committee (IMDMCC).
(b) A task force should be set up in the Ministry to guarantee that a working guideline and a risk reduction action plan are drafted.
(c) Make forestation efforts involving locals a major priority, with a focus on coastal regions, chars, and islands.
(d) Social forestation relies on strong assistance from a variety of local and national non-governmental organizations.

(e) Ensure adequate and effective conservation of the country's existing forests, and draft suitable legislation to prevent environmental deterioration caused by the release of toxic gases or liquids from the chemical industry.
(f) Make sure there's money in the budget for its risk mitigation efforts.
(g) Plan climate change studies and preparations.
(h) Mangrove afforestation and further growth of coastal and island forests are two examples of risk management and risk reduction activities that might get financing.
(i) Take control of environmental issues like pollution and climate change.
(j) Enact laws to prevent the release of harmful chemicals and other contaminants into the environment.
(k) Projects involving the planting of trees along highways and embankments should be prioritized in high-hazard regions.
(l) Assist local initiatives aimed at lowering risk, such as tree-planting campaigns.
(m) Oversee efforts to educate the public about environmental and safety risks.
(n) Create a strategy for reducing risks and being ready in each industry.
(o) Make preparations for a variety of emergency situations across different industries.
(p) Make sure the Ministry has a backup strategy in place.

Emergency Response

Normal Times

(a) Appoint one Ministry Liaison Officer to serve as the Department's main point of contact for disaster management.
(b) Give primary focus to increase forest coverage in coastal regions and on islands in the sea.
(c) Initiate mangrove afforestation in forest areas.
(d) Monitor the unfolding tragedy for any signs of environmental degradation to forestall the escalation of danger.
(e) Make that afforestation expansion programs are fully funded and put into action.
(f) Create a legislation to protect the environment against companies that produce chemicals or that release gases or liquids that pollute the atmosphere.

Alert and Warning Stage

(a) Send out a warning and instructions to the forest and environment departments so that they may take the appropriate measures to ensure the safety of people and property in the field.

Disaster Stage

(a) Establish communication with the EOC at the MoFDM and activate the Ministry's Control Room.
(b) Move quickly to remove uprooted trees from roadways.

Rehabilitation Stage

(a) Monitor the effects of disasters on the environment to ensure its protection and take action using the personnel and resources of the Forests and Environment Departments in concert with the government at all levels, the social sector, and non-governmental organizations.

Department of Forests

In addition to normal responsibilities, the Forest Department (FD) will perform the following duties.

Risk Reduction

(b) Help the Ministry of the Environment and the Forest draft laws and regulations to save the forest.
(c) To guarantee afforestation in coastal regions, chars, and islands, a set of guidelines incorporating locals should be drafted.
(d) Social forestation relies on strong assistance from a variety of local and national non-governmental organizations.

Emergency Response

Normal Times

(a) In cyclone-prone locations, it is important to plan and carry out extensive afforestation operations on highways and embankments.
(b) Mangrove reforestation should be prioritized in cyclone-prone regions.
(c) Community afforestation initiatives in disaster zones should be supported and encouraged.
(d) Staff members at the Department of Forests need to be educated on how to alert the public to impending disasters and how to save the lives of those in danger.

Alert and Warning Stage

(a) Direct your department's personnel in disaster zones to safeguard agency property and assist the affected community as best you can.
(b) Install a single Command Center in the department's main office.

Disaster Stage

(a) Get the word out to your department's officers and employees so they can support the people in the regions hit most by the tragedy.
(b) Clear the roadways of any fallen trees.

Rehabilitation Stage

(a) Determine the extent of forest property damage and report your findings to the IMDMCC and the Ministry of Environment and Forests.

Department of Environment

Besides normal functions, the Department of Environment will perform the following duties.

Risk Reduction

(a) Determine the potential risks, consequences, and adaptation strategies associated with natural disasters and climate change on both a national and local scale.
(b) Create a strategy for national adaptation and put it into effect.
(c) Develop and carry out localized efforts to preserve natural resources.
(d) Create a data repository of information on environmental changes on a national, regional, and local scale.
(e) Get involved in a variety of global talks.
(f) Make environmental impact assessments mandatory for all capital expenditures.
(g) Develop and roll out a variety of education and awareness campaigns aimed at raising everyone's level of comprehension.

Emergency Response

Normal Time

(a) Determine the most pressing environmental concerns in disaster-stricken regions and implement solutions.
(b) Provide guidance to local authorities on how to manage and avoid environmental damage in the aftermath of a catastrophe.

Alert and Warning Stage

Same as (a) and (b) above.

Disaster Stage

(a) Environments in places hit by disasters need close monitoring, and response should be prepared in advance.

Rehabilitation Stage

(b) Evaluate the effects on the environment and provide policymakers with recommendations.
(c) Advise and help the local government remove or control the causes of environmental degradation as soon as possible after a catastrophe, so that preventive measures may be taken.

Ministry of Information

The Ministry of Information can help the public cope with the situation with composure and resolve so that as little harm as possible is done as a result of the floods and storms. This Ministry will also aid government workers of all ranks in better understanding their roles in disaster preparation and responsiveness to the public.

The Secretary of the Ministry will perform the following duties:

Risk Reduction

(a) Appoint a high-ranking official in this Department to act as the Ministry's primary contact for disaster management.
(b) Find a way to assign a Ministry liaison to each department.
(c) The transmission of risk reduction information in the electronic and print media is encouraged via the issuing of directives to the radio, television, news media, press information department, mass communication department, and films & publishing department.
(d) Control and oversee the activities of the Radio and Television of Bangladesh, the News Media, the Press Information Department, the Mass Communication Department, and the Films and Publications Department.
(e) Through the use of television, radio, and other media, a widespread public may be reached and educated on what they can do to prepare themselves, their families, and their communities for an emergency and what they can do to help in the aftermath.
(f) Create a protocol for TV and radio that translates technical terms used in alerts into language that the general public can comprehend.
(g) Spread the word by widely disseminating cyclone and flood alerts.
(h) Make a Ministry emergency plan.
(i) Create a close working relationship with MoFDM.
(j) Make sure there's money set aside to manage measures taken to reduce risk.
(k) Make that the agency has a backup plan in place to protect people, machinery, and supplies in the event of an earthquake.
(l) Implement a wide-ranging media effort to educate the public on the need of being ready for an earthquake. (The Director of the GSB shall see to it that the public is informed about earthquake dangers in terms they can understand.)
(m) Spread the word about what people may do to protect their homes and families. The DMB Director General will supervise the dissemination of pertinent data.
(n) Arrange for widespread education on earthquake safety to ensure future buildings are compliant with the Building Code (Ministry of Public Works will support on this).
(o) Get the word out about what the government and its agencies are doing to provide immediate aid and long-term recovery to the impacted populace.

Emergency Response

Normal Times

(a) Make use of television, radio, and other forms of public relations to disseminate information provided by the Disaster Management Bureau and the relevant Ministry about disaster prevention, response, and recovery.
(b) Spread the word far and wide with the aid of the media about the cyclone and flood warning signs, and make sure everyone knows what they mean (Director, BMD will provide details of explanatory notes in simple and easily understandable language).
(c) Make sure that the Press Information Department, the Mass Communication Department, the Films and Publications Department, and Bangladesh Television

adhere to their mandates at all times, but particularly during warning and crisis situations.

(d) Ensure the safety of one's own property and infrastructure by taking all necessary precautions.
(e) Ensure these mass media have full assistance as needed to disseminate early warnings.
(f) Instruct all of its departments to help the administration at the District, Upazila, and UP levels.

Alert and Warning Stage

(a) Establish a Ministry Control Room and appoint a Liaison Officer responsible for communicating with the MoFDM's Control Room.
(b) Spread clear information about the responsibilities of the people living in the danger zones, and make sure that preventative and warning signals are widely publicized as recommended by the BMD.
(c) In the event of a cyclone, it is imperative that the most up-to-date meteorological reports be widely disseminated.
(d) Plan for the rapid collecting and distribution of weather bulletins to preserve their value as news.
(e) Do everything you need to in order to safeguard your possessions.

Disaster Stage

(a) Take shifts in the Control Room (24 h).
(b) Make sure the information being shared with the public is an accurate depiction of the situation, one that will calm rather than frighten listeners.
(c) Proceed with the NDMC and IMDMCC-requested public announcement of situation-related news and orders.
(d) Facilitate the travel of local and international journalists to the affected region so that a fair and balanced report may be published.

Rehabilitation Stage

(a) Help get the word out about the many ministries, departments, and agencies working on relief and rehabilitation for the impacted population, both in the immediate and long terms.

Bangladesh Betar (Radio Bangladesh)

Bangladesh Betar plays a crucial role before, during, and after a natural catastrophe. Following directives from the government, Bangladesh Betar would provide warnings, inspire and educate the populace, and provide details on how to prepare for and respond to disasters. Bangladesh Betar will be responsible for the following in the event of a natural disaster:

Risk Reduction

(b) Connect with the BMD, FFWC, MoFDM, and the T&T board by phone, fax, and email.

(c) Collaborate with the BMD and MoFDM to create awareness and public education programs that include topics such as warning signs and their significance, how to avoid, prepare for, and respond to disasters on an individual, family, and community level.
(d) Disseminate briefings on how the government approaches catastrophe risk management.
(e) Spread the word about the updated cyclone warning signals at river and seaport events.
(f) Make that the agency has a backup plan in place to protect people, machinery, and supplies in the event of an earthquake.
(g) Share the information about how people may protect themselves and their families at home.
(h) Get the word out through radio to the general public.
(i) In order to reduce the danger posed by earthquakes, you need to have the right information disseminated.
(j) Set up government-run media broadcasts. Emergency procedures for dealing with earthquakes; instructions/standing orders.
(k) Send encouraging programming to the regions where it is needed most.
(l) Life safety, rescue, securing/reserving home items, and water storage should all be included prominently in any emergency preparedness broadcast.

Emergency Response

Normal Times

(a) Spread the word about a unique initiative to have people ready for earthquakes and other disasters at home.
(b) Set up the fastest possible means of contact between the BMD, the Flood Forecasting and Warning Centre, and the T&T Board, and keep those lines of communication open at all times.
(c) Set up a faxing system that allows for constant contact with DMB and BMD (and email).
(d) To educate the public and encourage participation in disaster preparation, mitigation, and preventive efforts, it is important to share specific plans for action with BMD and the MoFDM.
(e) Promote warning signal interpretation programs in tandem with the BMD, DMB, and CPP.
(f) When airing special weather bulletins, be sure to include an audio briefing from the Storm Warning Centre's (SWC) cyclone forecaster based on weather charts, radar, and satellite images.
(g) Maintain a backup plan to keep broadcasting around the clock if required.

Alert Stage

(a) At predetermined intervals, have all broadcasting stations broadcast the weather report after receiving warning signals 1, 2, or 3 from the Meteorological Department.

Warning Stage

(b) As soon as you get warning signal No. 4, make sure it is broadcasted every hour along with the explanations you get from the BMD, and if the MoFDM tells you to keep on, you should.
(c) The DM&RD MoFDM announcement was broadcast on a local radio station in Dhaka. Chittagong, Rajshahi, Sylhet, Rangpur, and Khulna Radio Stations will broadcast local authorities directives if necessary. The program would benefit from using the native language.
(d) As soon as you get the Flood Warning, be sure to broadcast it.
(e) It is recommended that, if instructed to do so by the Ministry of Food and Disaster Management, the danger signal be aired every 30 min and the great danger signal be broadcast every 15 min during a cyclone.
(f) If directed by the MoFDM, notify the local administration/authority of the evacuation order for the people residing in susceptible regions.

Disaster Stage

(a) Per the Ministry of Foreign and Domestic Affairs' recommendations, during a cyclone the danger signal should be issued every 30 min, and the great danger signal every 15 min.
(b) Disseminate information on how to be safe in the case of a natural catastrophe.
(c) Broadcast safety programs covering issues including survival, rescue, household items, drinking water, health, etc.

Rehabilitation Stage

(a) Play shows that would boost people's spirits in the impacted areas.
(b) Rehabilitate both the short and long term by spreading the word.

Bangladesh Television

Following a calamity, Bangladesh Television will broadcast the following coverage.

Risk Reduction

(a) Make sure your phone and teleprinter are working properly by getting in touch with BMD and, if required, setting up an appointment to use the T&T Board's non-exchange Magneto phone line. Establish continual fax (email) contact with BMD, FFWC, and DMB.
(b) Broadcast special TV shows on multiple channels focusing on risks, risk reduction, risk treatment strategies, response and recovery coordination, public safety, and preparedness activities at the individual, family, and community levels, such as talk shows, debate competitions, short films, and teledrama.
(c) Jointly produced by the BMD, DMB, and MoFDM, air special education and awareness programs on television.
(d) Show documentaries and short films about catastrophe preparedness.
(e) Get the agency's backup plan together to protect the employees and the equipment.

(f) Construct a transmission tower and associated equipment if necessary to reduce exposure to seismic hazards.
(g) Show documentaries and other programs that educate the public on how to prepare for and deal with earthquakes.
(h) The government should coordinate the transmission of all emergency announcements.
(i) Broadcast critical details on how to lessen earthquake dangers, such as how to implement construction regulations correctly.
(j) Set up government-run media broadcasts. Emergency procedures for dealing with earthquakes; instructions/standing orders.

Emergency Response

Normal Times

(a) Broadcast special programs for the public's education and mobilization as approved by the BMD and the MoFDM.
(b) Broadcast the Meteorological Department's specific warning signals and explain what they signify.
(c) While broadcasting these rare weather alerts, be careful to include a voiceover briefing from a cyclone forecaster at the Storm Warning Centre (SWC) and visuals of radar and satellite imagery.
(d) Play videos and short films on what to do in the event of a natural catastrophe such as flood, hurricane, or earthquake.

Alert Stage

Bangladesh Television shall promptly air signal Nos. 1, 2, and 3 from Dhaka Centre and all Sub-Centres upon receipt from the Meteorological Department and will continue to do so at regular intervals afterward. Flood alerts generated by the Flood Forecasting and Warning Centre should also be broadcast.

Warning Stage

(a) Bangladesh Television shall transmit on receipt of warning signal No. 4 along with explanation provided by Meteorological Department every one hour and continue such telecast if instructed by the Ministry of Food and Disaster Management without any gap even beyond regular broadcasting hours. As soon as signal No. 3 is hoisted, Bangladesh Television will make communication with the Ministry of Food and Disaster Management on full-time basis for nonstop transmission beyond usual broadcasting period.
(b) Ensure audio briefings from the cyclone forecaster (SWC) and visual depiction of radar and satellite photos when telecasting special weather bulletins.
(c) All broadcasts from the Meteorological Department and the Ministry of Food and Disaster Management will be broadcast on Dhaka Television.
(d) Avert disaster by spreading the word about the Meteorological Service's and the Ministry of Food's preventative actions.

(e) When receiving a flood warning, broadcast it on television in great detail.

Disaster Stage

(a) The Ministry of Food and Disaster Management recommends broadcasting cyclone alerts every 30 min and the Great Danger Signal every 15 min, even outside of regular broadcasting hours.
(b) Broadcast the directives given by the Ministry of Food and Disaster Management to local administration/authority to relocate people from dangerous regions.
(c) Broadcast lifesaving tools and rescue and property and water protection tips.

Rehabilitation Stage

(a) Telecast programming to keep the morale of the affected people high.
(b) Telecast short- and long-term programming on recovery.

Mass Communication Department

The Mass Communication Department shall perform the following duties regarding disaster:

Risk Reduction

(a) Encourage the use of videos, films, pamphlets, and other forms of mass communication to raise public awareness, especially of risks, risk reduction, risk treatment techniques, response and recovery coordination, public safety, and preparatory actions at the individual, family, and community levels.
(b) Instructions for maintaining physical infrastructure should be made public in order to minimize the amount of damage and loss.
(c) Produce video, cinema, film, slide, pamphlets, and pot songs on earthquake disaster management in order to raise public awareness on national and local levels.
(d) Arrange publicity for the follow-up of the Building for seismic risk reduction.
(e) Show a movie at the theater on quake safety precautions.

Emergency Response

Normal Times

(a) The following are some of the things the Mass Communication Department will do using video, cinema, films, slides, pamphlets, etc., to raise catastrophe awareness among the general public.
(b) Measures should be implemented in normal times for the reduction/mitigation/preparedness of catastrophe useful for decreasing loss and damage during disaster.
(c) Emergency response duties and obligations in times of crisis.
(d) Publicize guidelines for physical infrastructure to prevent loss and damage from the catastrophe.
(e) Consult the DMB for expert guidance on these technical matters.

Alert and Warning Disaster Stage

(a) Using a variety of public relations tools, keep the people in the affected regions aware of what they can do to help.

Rehabilitation Stage

(b) At this point, it's important to do promotional work in the afflicted region to keep people's spirits up and get life back to normal. Help people prepare for and recover from epidemics, natural disasters, and other emergencies by providing tailored guidance on topics including self-reliance in rebuilding, general security, agricultural rehabilitation, and more.

Press Information Department (PID)

The PID shall perform the following duties regarding disaster management.

(c) Maintain a consistent media effort to raise awareness of potential dangers, risks, and catastrophes by publishing a range of articles, pamphlets, and official press releases.
(d) Make sure a campaign is launched to teach people about the need of being prepared for catastrophe information.
(e) Information on disasters should be gathered and shared with both international and national media outlets.
(f) Give the media the facts so there won't be any fear caused by unfounded stories.

Department of Films and Publications

The department shall perform the following duties regarding disaster management.

(a) Create and refine a range of media (video, tell-of slogans, communication messaging) to support catastrophe management initiatives across a spectrum of risks.
(b) Create a variety of documentaries and videos based on lessons learned in the field to inspire people to better prepare for emergencies.

Ministry of Posts and Telecommunications

Risk Reduction

(a) Establish communication with disaster management committees and DM&RD, and name a high-ranking official to serve as the Ministry's Disaster Management Focal Point.
(b) Make a backup plan for the Ministry's and any related communication systems.
(c) Connect the Tsunami Early Warning System in the Indian Ocean with the rest of the world through direct communication.
(d) Establish communication connection to worldwide network to aid in cyclone and other forecasting efforts.

Bangladesh Telecommunication Regulatory Commission (BTRC)

Risk Reduction

(a) Set up communication with the National Disaster Management Council (NDMC), the International Disaster Management Cooperation Council (IMDMCC), and the National Disaster Management Assistance Council (NDMAC).
(b) Send out orders to the Divisional Commissioner, DC, TNO, and UP-Chairperson to assist the NDMCC, IMDMCC, and NDMAC as needed.
(c) In order to quickly disseminate early warning information to the community, it is important to set up networks with all of the mobile phone carriers.
(d) Direct the planned assessments of danger and measures taken to lessen it, so that essential services and infrastructure, such as the mail and post, government documents, currency, and other assets, remain undamaged and unlost.
(e) Have the appropriate people take the necessary precautions to ensure that telephone services are available around the clock before it becomes an emergency.
(f) Help the MoFDM and BMD set up a reliable early warning system.
(g) Conduct a sector-specific risk analysis and build a sector-specific strategy to deal with potential threats.
(h) Make sure money is set out in the budget for risk mitigation efforts.
(i) Get in close touch with MoFDM and start working together.
(j) Providers of communications services should be instructed to adhere to the guidelines and offer the necessary backup services in the event of a catastrophe.
(k) Management and technical personnel should band together to form a task force that can function in an emergency.
(l) Increase the task force's preparedness for dealing with and responding to disasters by providing them with high-quality training.
(m) Collaborate across divisions to create a backup plan for handling potential crises.
(n) If required, it should reorganize its resources, including its personnel.

Emergency Response

Normal Times

(a) In the event of a disaster, one Liaison Officer should be named as the focal point for management.
(b) In disaster-prone places, it's important to have a backup plan for communications, transportation, and supplies.
(c) Efforts to safeguard critical documents, currency, and postal items including stamps, seals, savings certificates, and government data against theft or destruction must be made a priority.

Alert and Warning Stage

(a) Allow access to the Control Room and run the machinery.

(b) In order to maintain communication between the Emergency Operations Center (EOC) and the Ministry of Food and Disaster Management and the Disaster Management Bureau, a Liaison Officer should be appointed.
(c) In the event of an impending crisis, emergency frequency allocation must be made for wireless communication.
(d) Get the word out fast to designated locations by text message, email, voice mail, and other electronic means.
(e) Connect all populated islands through Wi-Fi.

Disaster Stage

(a) Assume responsibility for the day-to-day upkeep of your telephone system.
(b) Make that the emergency phones and teleprinters of government departments, nonprofits, and organizations like the Red Cross and the Center for Peace and Prosperity (CPP) are always charged and ready to go. When receiving complaints from the Control Rooms of Ministries involved in disaster preparation and emergency activities, or similar organizations, swift action must be taken.
(c) Maintain communication with the service providers and provide any necessary assistance.
(d) It is important to have a backup plan for communication in place in case of an emergency and to have it available for use in the impacted regions.

Rehabilitation Stage

(a) In the event of a breakdown in telephone or telegraph service, an emergency crew should be on standby to make repairs and reinstall any necessary equipment.
(b) Connect the necessary equipment for emergency telephone, telegraph, and wireless broadcasting.

Bangladesh Telecommunication Company Limited (BTCL)

In addition to his regular responsibilities as Chairman of BTCL and the execution of the company's work plan (contingency plan), he will also be responsible for the following.

Risk Reduction

(a) Assess the risks and vulnerabilities of certain sectors and create a strategy for mitigating those risks.
(b) Follow through with the strategy.

Emergency Response

Normal Times

(a) Choose one BTCL Liaison Officer to serve as the organization's main point of contact during times of crisis.
(b) The BTCL Headquarters and regional offices should function as a central information hub for affected communities.

(c) Prepare for the safety of your own personnel, infrastructure, and property in disaster zones.

Alert and Warning Stage

(a) In disaster-prone locations, you should make preparations for access to emergency telephone services.
(b) Maintain a team of mechanical engineers, technicians, and anybody else in charge of telephone and telegraph line installation at the ready at all times with the necessary tools for impromptu repairs and rebuilding.

Disaster Stage

(a) You should act quickly and efficiently to ensure that telephone services are available around the clock.
(b) Maintain the functionality of all Ministry, Division, and Directorate emergency telephones and teleprinters, as well as those of other agencies providing emergency services, such as the Red Cross, the Civil Protection and Preparedness Agency, and other similar organizations.
(c) Get right on fixing the issues being raised by the ministries and agencies in charge of cyclone preparation and lifesaving programming's control rooms.
(d) Keep communication equipment on hand and ready for distribution in disaster zones.
(e) If the Armed Forces Division is ever sent somewhere far away, they'll need a way to get in touch with home.

Rehabilitation Stage

(a) Help restore communication services as soon as possible after a tragedy. You should take extra precautions with telecommunications infrastructure on coastal islands.
(b) Plan and submit proposals to the Planning Commission for approval of funding to restore damaged telephone and teleprinter infrastructure.

Director General, Bangladesh Post Offices

Director General shall carry out the following responsibilities in addition to routine operations and the execution of its own Action Plans (contingency plan).

Risk Reduction

(a) Assess the sector's risk and exposure, pinpoint the weak building, then develop and execute a strategy to fortify it.
(b) Locate secure areas where new buildings may be built without risk.

Emergency Response

Normal Times

(a) Choose one Directorate Liaison Officer to serve as the group's primary point of contact for disaster management.

(b) Prepare your own staff in advance of disasters by providing them with training on the highest priorities that must be met in affected regions.
(c) The EOC of the DM&RD, MoFDM, needs one Liaison Officer to maintain communication with the field offices and Control Room of the Ministry of Posts and Telegraphs.
(d) Every year, in the months of April and September, conduct rehearsals/drills in disaster zones.

Alert and Warning Stage

(a) In places where cyclones or floods are predicted, take precautions to ensure the safety of department employees, facilities, and property.

Disaster Stage

(b) Utilize subordinate authorities and others as needed to implement effective measures for protecting post and telegraph services.
(c) Keep the telephone and post offices open in disaster zones.
(d) Prevent loss or theft of mail, government documents, cash, and precious items like postage stamps, seals, savings certificates, critical records, etc. by using appropriate security measures.

Rehabilitation Stage

(a) If it becomes essential, temporary post offices should be set up to serve those who have returned from a disaster shelter or relief center to their homes.
(b) Help local authorities maintain postal services while they deal with evacuees and conduct rescue, aid, and recovery activities.
(c) Send out directives to all parties involved, outlining the need of maintaining postal services during times of emergency.

Ministry of Local Government, Rural Development, and Cooperatives

Risk Reduction

(a) Conducting a thorough review of sectoral risks can help you create a strategy to deal with them.
(b) It is important to set up a system of monitoring and assessment and provide sufficient funds for the plan's execution.
(c) Involve local governments in assessing catastrophe risk in order to better plan for and build local transportation infrastructure including roads, bridges, and culverts.
(d) Help the government build helipads on remote islands that can withstand natural disasters.
(e) Promote educational and awareness drives to get people to take preventative measures, such as stocking up on supplies, before storm and cyclone season hits.
(f) Get the Ministry to set up a framework for communicating risks across different industries.

(g) Make sure the agency has a backup plan in place, and make sure that plan is up to date.
(h) Adhere to the BNBC and make any adjustments that are needed to make sure it's carried out as intended.
(i) Training programs for government engineers, planners, and architects on infrastructure and urban planning may help reduce the impact of earthquakes on building and urban development.
(j) Collaborate with the GSB to determine potential seismic hazards and guarantee that expert engineers are involved in the rebuilding of damaged regions.

Emergency Response

Normal Times

(a) Identify and plot watering holes, reservoirs, and other sources of potable water for the local population.
(b) Assemble public health engineering units, stock up on emergency supplies, and manage reserve inventories of tube wells as part of your disaster preparations.
(c) Be sure that schools, high ground, reinforced earthen mounds, and private structures that might be utilized as shelters in high-risk locations are in good condition.

Disaster Stage

(a) Establish and maintain regional disaster management coordination committees.
(b) Protect women and children throughout the population evacuation process.
(c) Safe drinking water must be made available in the impacted communities, which may require immediate repairs to broken tube wells.
(d) Collaborate with the Ministries of Women and Children Affairs and Social Welfare to organize a swift evaluation of damage and vulnerability of persons, paying particular attention to women and children (orphaned, separated, displaced).

Rehabilitation Stage

(a) Quickly evaluate the extent of the damage, losses, and rebuilding required, and act accordingly.
(b) Coordinate a community-driven approach to home repair.
(c) Schedule tube well repairs and resinking with the DPHE.
(d) Fix or replace damaged stretches of pavement, drainage structures, and pedestrian bridges.
(e) When dealing with a crisis, it's important to keep everyone informed and updated as you work to recover.

Local Government Division (LGD)

LGD will carry out the following responsibilities in addition to its regular operations and the execution of its own Action Plans.

Risk Reduction

(a) You should name one LGD Liaison Officer as the group's primary contact for disaster management.
(b) Make sure that earthquake risk, among other existing and potential hazards, is included into the design of every building.

Emergency Response

Normal Times

(a) In order to facilitate travel to cyclone shelters and development hubs, it is imperative that local government agencies invest in the construction of roads, bridges, and culverts.
(b) Coastal Union Parishad complexes, Upazila Headquarters, and offshore islands should all have clay mounds and helipads built to protect them.
(c) I would advise folks living in cyclone-prone locations to build at least a single-room brick house. Encourage residents to raise the level of their home's foundations above the expected level of floods and storm surges.
(d) Union Parishads should encourage collaboration between various community groups and organizations, such as the Village Defense Party, Ansar, Family Planning Worker, Agriculture/Livestock/Fishing/Education Department employee, union social workers, community youth group, NGOs, and CPP volunteers.
(e) Safeguard rescue equipment at the Union Parishad Headquarters and emergency supplies at the Upazila level.
(f) Form the appropriate engineering units for public health on all levels.
(g) Create plans highlighting high-density areas of people as well as watering holes and safe havens.
(h) Maintain a safety store of spare components and tube wells.

Alert and Warning Stage

(a) Get the Local Government Control Room up and running.
(b) One Liaison Officer should be appointed to serve as a point of contact between the EOC and the MoFDM and DMB.
(c) Warn everyone you know that the cyclone/flood tragedy is imminent.
(d) Organize a rescue, relief, and rehabilitation committee and a group of volunteers from the local community.
(e) Collaborate with relevant Ministries and aid organizations to ensure smooth operations.
(f) In cyclone and flood-prone locations, it is important to provide shelters, reinforced earthen mounds, high lands, private buildings, schools, and secure high lands to protect people and cattle.

Disaster Stage

(a) Maintain Control Room accessibility throughout the clock (24 h).
(b) Maintain the functionality of the Disaster Management Coordination Committees.

(c) As soon as you get instructions to do so, make sure the populace is evacuated.
(d) Recruit one Upazila-level official per cyclone shelter to handle paperwork.
(e) Maintain a reliable supply of potable water in case it is required.
(f) Maintain constant communication with the local government.
(g) Instruct the Upazila Nirbahi Officer and the Deputy Commissioner to reach out to local government entities for help with rescue and relief efforts.
(h) Make sure there are people available to make emergency repairs to broken tube wells in all impacted Upazilas. Technicians from safe regions should be sent to assist those in need.

Rehabilitation Stage

(a) Assess the level of harm or loss.
(b) Participate in and aid in any and all rescue and relief efforts.
(c) Rebuilding damaged homes should be coordinated on a self-help basis using local resources and government-provided supplies.
(d) If the tube wells need fixing or sinking again, contact the DPHE about setting up an appointment.
(e) Initiate the process of repairing or rebuilding cyclone- and flood-damaged minor roads, culverts, and bridges.
(f) Stable prices for food, water, and other necessities can only be achieved via the joint efforts of the Union Parishad, Upazila Parishad, Municipal Corporation, and Town Committees.
(g) In the aftermath of a cyclone, the Department of Public Health and Environment (DPHE) should be tasked with drawing up plans for the installation of deep and shallow tube wells in the affected districts.
(h) Funding and plans for minor roads, bridges, and culverts in cyclone/flood-prone regions, as well as community centers that may serve as temporary shelters, should be prepared by the LGED.

Local Government Engineering Department (LGED)

The LGED and Engineers will be accountable for the following tasks in their respective domains, each with the authority delegated to them by the LGED or the Engineers.

Risk Reduction

(a) Choose one LGED Liaison Officer to serve as the group's primary point of contact for disaster management.
(b) The LGED's Action Plan should take into account the full scope of potential dangers.
(c) Plan and build feeder roads, bridges, and culverts to facilitate the free flow of water and the swift removal of flood waters.
(d) Union Parishads may be used to encourage and motivate individuals to build multi-story structures. In the event of a storm or flood, it is recommended that residents take refuge in at least one of the rooms on the top floors of brick homes that have been reinforced to withstand such events.

(e) Make maps showing major cities, watering holes (wells), water reservoirs, tube wells, and any other sources of potable water.
(f) Make sure the agency has a backup plan, and keep it up to date.
(g) Adhere to the Building Code and take the appropriate precautions to verify that it is being followed whenever this is required.
(h) Training programs for government engineers, planners, and architects on infrastructure and urban planning may help reduce the impact of earthquakes on building and urban development.
(i) Collaborate with the GSB to determine potential seismic hazards and guarantee that expert engineers are involved in the rebuilding of damaged regions.

Emergency Response

Normal Times

(a) Keep the roads, bridges, and culverts that lead to cyclone shelters, schools, and hospitals in good working order so that people can get to them quickly and safely in case of an emergency.
(b) In cyclone-prone locations, experts recommend elevating water reservoir (pond) banks beyond the level of cyclone-induced tidal bore and in other places above flood level to protect residents' access to safe drinking water and animals' ability to find refuge. It would be ideal to plant a number of parallel rows of trees.
(c) Keep a supply of Baily bridges on hand for mending damaged bridges and culverts, as well as for patching up roads so that emergency supplies can travel quickly and easily.
(d) Upazila Parishad-managed storm shelters should have their foundations raised above the floodplain and maintained on a regular basis.
(e) Engineer, Union Parishad Secretary, and other technical and non-technical staff members should be educated on and prepared to disseminate information on emergency management and disaster preparedness programs.

Alert and Warning Stage

(a) Run a single Control Room out of the Directorate's main office.
(b) Assist in evacuation, rescue, and relief efforts by attending the meeting of the local DMC.
(c) Please spread the word about the impending tragedy and the need to take preventative measures to save supplies, stock, and tools.
(d) Take the necessary measures to unclog the waterways in residential areas.
(e) Assist in locating safe sites in disaster-stricken communities, such as reinforced earthen mounds, private houses, schools, and Madrasahs, and high ground.

Disaster Stage

(a) Help with rescue, evacuation, and relief efforts by keeping the Control Room running nonstop (24/7).
(b) Collaborate with the Inter-Ministerial Committee for Disaster Management Coordination.

(c) Maintain lines of contact with emergency shelters and begin fixing any damage right once.

Rehabilitation Stage

(a) Determine the extent of the damage, loss, and requirements.
(b) Help with any and all rescue and recovery efforts.
(c) Rebuilding damaged homes, roads, bridges, and culverts should be coordinated on a self-help basis using local resources and government supplies.
(d) Pick up the pieces and start rebuilding the tiny roads, bridges, and culverts that were destroyed by the calamity.
(e) Plan both immediate and long-term maintenance and construction of critical backroads for use in communication, evacuation, and disaster relief.

Rural Development and Cooperatives Division (Including Bangladesh Rural Development Board)

In addition to carrying out its mandated responsibilities and executing its own work plan, the Rural Development and Cooperatives Division/Bangladesh Rural Development Board will be responsible for the following.

Risk Reduction

(a) One officer should be designated as the liaison and the point person for disaster management.
(b) When developing the BRDB's Action Plan, it is important to take into account the whole range of potential hazards.
(c) Implement educational initiatives aimed at empowering small- and medium-sized livestock and poultry producers.
(d) Aid efforts to raise public awareness about the need of risk analysis and mitigation education.

Emergency Response

Normal Times

(a) It is important to establish a contingency fund in the event of a catastrophe, which may be used to quickly address the following issues.
(b) Use TCCA and KSS to aid in rural reconstruction after natural disasters.
(c) Using the principles of mutual aid and self-help, arrange for TCCA to provide specialized training classes for cooperative members on topics such as building homes and farming land.

Alert and Warning Stage

(a) Run the Operations Center at BRDB HQ, TCCA, ATCCA, and BRDB Field Offices, and work with local authorities at all levels to put post-disaster recovery plans into action.
(b) Assist in disaster relief efforts such as evacuation and search and rescue by attending meetings of the local disaster management committee.

(c) In the event of an impending catastrophe, the executive staff of BRDB, ATCCA, and TCCA should notify all employees and take appropriate steps to secure the safety of all storage facilities, inventory, and tools.

Disaster Stage

(a) Assist the local government via TCCA in locating safe havens for people and animals, move TCCA stock to a more secure location while keeping the local government informed, and enlist cooperative society members as part of the evacuation team.
(b) If you discover casualties, property damage, and the need for aid and rehabilitation, you must immediately notify the appropriate Ministry or Office.

Rehabilitation Stage

(a) Find out how much money cooperative members lost, and then, after you know-how much money they need for production loans, distribute the money to them fairly.
(b) Assist in determining the need for various inputs, planning production schedules, and acquiring these things from multiple organizations.
(c) Officials must prepare production plans for primary agri-cooperative societies and other societies, and loans must be quickly approved and disbursed following consolidation of demand.
(d) Prepare loans to fulfill the needs of those whose lives have been upended by the storm.
(e) Make individualized plans for the recovery of each demographic.
(f) In the affected regions, set up special programs for acquiring the necessary components for constructing new and repairing existing irrigation tube wells.
(g) Gather the populace and non-governmental organizations (NGOs) to work together on development plans to repair the harm.
(h) It is important to rally the affected farmers together and persuade them to implement cooperative agricultural production and distribution plans.
(i) Make sure that authorities coordinate BBS and MBBS programs for low-income persons with non-governmental organizations (NGOs), and that training and debt consolidation are done so that these people may find gainful employment.
(j) On every level, the Cooperative Directorate will collaborate closely with the BRDB.

Department of Public Health Engineering (DPHE)

Due to arsenic pollution in groundwater and also salinization in surface water in the coastal area, ensuring clean drinking water has become a major concern for the DPHE. When individuals are unable to access clean water sources during a crisis, they often resort to using dirty water, leading to an outbreak of diarrhoeal illness. In addition to their regular responsibilities, DPHE should also be responsible for the following.

Risk Reduction

(a) Consult with the local community to determine what threats they face, especially in regard to access to clean water and good hygiene habits, and then plan how to mitigate those threats.
(b) Formulate a coordinated strategy for enlisting the help of public and non-profit organizations, as well as the commercial sector, to carry out the plan for mitigating risks.
(c) Create synergies and reduce duplication by coordinating the efforts of non-governmental organizations and other private sector entities.
(d) Give them the help they need to keep things clean and sanitary.
(e) Create a system of checking in and checking out to make sure the money is being put to good use.

Emergency Response

Normal Times

(a) Find potential catastrophe zones and make sure there are enough tube wells sunk to provide a steady supply of clean water.
(b) Tube wells are vulnerable to damage from natural disasters such as tidal bores and floods, and therefore it's important to have replacement components on hand.
(c) Safeguard an adequate supply of bleach powder for use in emergency situations.
(d) Prepare a technical or repair crew to work in cyclone and flood zones.
(e) Every six months, you should check to make sure you have enough bleaching powder and replacement components on hand.
(f) During times of crisis, it is important to promote the building of water-resistant latrines that are both inexpensive and sanitary.
(g) Save extra tube wells and water-resistant latrines for use in places without them or in emergency shelters after a natural disaster.
(h) Make sure you have enough cash on hand to cover any unexpected costs that may arise.
(i) People who may be at risk should be taught how to properly utilize water purification tablets (WPT) and beaching powders.

Alert and Warning Stage

(a) Assemble and station tube well repair teams in preparation for deployment to disaster zones.
(b) In disaster-prone regions, assess the availability of tube well replacement components and, if required, make preparations to stock up.

Disaster Stage

(a) To fix the damage caused to tube wells and water supply lines by tidal bore and flood waters, dispatch repair crews to the affected locations.
(b) If regular water supply has been cut off in a region, make arrangements for emergency water delivery (man can live for longer period without food than without water).

Rehabilitation Stage

In coordination with the DMC and on orders of the Committee.

(a) Keeping the water flowing for consumption is a priority.
(b) Provide a sufficient amount of bleaching powder so that shelters, relief camps, etc., can maintain a clean and sanitary environment.
(c) See to it that the necessary resources (both human and material) are on hand to get the tube well and water supply system back up and running as soon as possible.

Dhaka WASA

Dhaka WASA will do the following in addition to its regular responsibilities and the proper execution of agency operations.

Risk Reduction

(a) Appoint a single trusted WASA employee as the organization's official "Focal Point for Disaster Management."
(b) Participate in all meetings of the Disaster Management Committee, from the National level on down to the Upazila level, to coordinate efforts for relief, evacuation, and reconstruction.
(c) Start planning for an alternate water supply and sewage and drainage system by conducting a sectoral risk assessment effort.
(d) Develop a backup strategy for WASA in the event of an earthquake or other natural calamity.
(e) Set up WASA employee with training to better serve the agency during and after disasters.

Ministry of Housing and Public Works

The Secretary of the Ministry of Housing and Public Works shall guarantee the most efficient execution of the following in addition to usual tasks and the proper implementation of the Ministry's work plans.

Risk Reduction

(a) Appoint a single competent official within the Ministry to act as the Department's primary contact for disaster management.
(b) Participate in all meetings of the Disaster Management Committee, from the National level on down to the Upazila level, to coordinate efforts for relief, evacuation, and reconstruction.
(c) In order to formulate a strategy, it is necessary to conduct a sector-wide risk assessment.
(d) Directives should be issued to guarantee the National Building Code of Bangladesh is followed (BNBC).

(e) Prepare plans and strategies to pay for the restoration and renovation of Government buildings, buildings, systems, and infrastructure in high-risk regions.
(f) Involve post-disaster effect and risk assessment concerns in any rebuilding efforts.
(g) Create and roll out a program to educate employees at all levels on how to mitigate the effects of natural disasters on the housing and public works industries.
(h) Put in place a rigorous mechanism for checking the quality of public and private construction projects.
(i) Make sure the Ministry as a whole is ready for any emergency that might affect any of its departments or outposts.
(j) Make sure there's money in the budget for actions big and small.
(k) Initiate an interdepartmental and cross-ministry communication mechanism aimed at reducing risks.
(l) Make sure the agency has a backup plan, and keep it up to date.
(m) Make sure your BNBC for an earthquake is fully formed and ready to go in case of an emergency.
(n) Training programs for government engineers, planners, and architects on infrastructure and urban planning may help reduce the impact of earthquakes on building and urban development.
(o) Help the Geological Survey of Bangladesh pinpoint potential earthquake damage and get expert engineers involved in rebuilding efforts.

Emergency Response

Normal Times

(a) Appoint one Ministry Liaison Officer to serve as the Department's main point of contact for disaster management.
(b) Take part in DMCs at all levels, from the national to the Upazila, in order to help with relief efforts.
(c) To avoid any potential loss or damage during a catastrophe, the government should take the necessary precautions.
(d) Collaborate with relevant Ministries and local authorities to coordinate disaster management efforts.
(e) Send out orders to every department and field office, telling them to work with the Ministry to create a strategy for dealing with natural disasters.
(f) Put out an order for the government to take safety precautions. During a natural catastrophe, it is important to secure your property in order to prevent loss and damage.

Alert and Warning Stage

(a) Dole out some cautious advice to everyone involved.
(b) Prepare the Ministry's Command Center.

(c) One Liaison Officer should be assigned to liaise between the DMB and the MoFDM's EOC.
(d) Direct all District and Upazila authorities to ensure enough resources (both human and material) are available to safeguard and maintain public buildings and facilities.
(e) Move government supplies to a secure location if required to guarantee their safety.
(f) In the event of an emergency, it is imperative that you instruct the Directorate of Public Works to dispatch the appropriate personnel and resources from other locations to the impacted areas.
(g) Maintain communication with the municipal government and organize all program operations.

Disaster Stage

(a) Locate catastrophe zones, paying specific attention to Upazilas that may be hit hard.
(b) Safeguard all government supplies, equipment, and buildings.
(c) Make sure there are enough people and supplies to send to the impacted regions and to secure and restore government buildings.
(d) Instruct the Public Works Directorate to deploy more personnel and supplies to the affected regions to aid in emergency, relief, and recovery efforts, and to fortify and restore government property and infrastructure that has been damaged.
(e) In order to save life and documentation, orders must be issued to evacuate to a safe location.
(f) Assign personnel to the Control Room, where they may monitor the situation while simultaneously communicating with the DMB and the Emergency Operations Center (EOC) at the MoFDM.
(g) In the event that emergency reinforcement and repair work must be undertaken, you are thus directed to instruct the PWD to do so.

Rehabilitation Stage

(a) Compile a loss and damage report, detailing the extent of the harm and the extent to which it will need reconstruction or repair.
(b) Carry out a post-hazard impact analysis, and choose a course of action to lessen the negative effects.
(c) If similar calamities occur in the future, the government should have a plan in place to safeguard its property, and you should submit a thorough program with cost estimates to the relevant authorities.
(d) Make and keep safe financial provisions for the Ministry's urgent needs in catastrophe risk reduction.
(e) Make sure money is quickly allocated to where it can do the most good in fulfilling the needs of any rebuilding activities.

(f) Prepare precise work plans, schedules, and cost estimates for immediate and longer-term reconstruction of government property in order to repair, rehabilitate, and rebuild it.
(g) Make sure that all rehabilitation programs are coordinated with the other relevant Ministries and local authorities.
(h) As requested, you shall provide technical aid and manage the rebuilding project.

Public Works Department (PWD)

The PWD will carry out the following functions at Headquarters and via its subordinate offices at the Division, District, and Upazila levels in accordance with the delegation of powers by the department.

Risk Reduction

(a) The BNBC must be carried out correctly, so be careful to watch over it.
(b) All of the Department's development projects should take into account the potential impact of future disasters while formulating their policies, programs, and guidelines.
(c) Draft a guide on methods for assessing building stability and designing against earthquakes.
(d) Create and publish an updated list of susceptible buildings at regular intervals.
(e) Disseminate the Technical knowledge connected to earthquake and Tsunami to engineers.
(f) Back the efforts to retrofit.

Emergency Response

Normal Times

(a) To guarantee the correct implementation of the Building Codes, it is important to organize training programs and awareness campaigns with the civil engineers participating in the building works and the masons.
(b) Put in place a rigorous mechanism for checking the quality of public and private construction projects.

Alert and Warning Stage

(a) Set up a Control Room to coordinate with regional offices and the emergency response team.
(b) Once warning signs are received, an immediate alert must be sent to all parties involved.
(c) Maintain a ready force of guards and supplies to safeguard and fix up public buildings.
(d) If required, move all government supplies to a secure location to guarantee their safety.
(e) Send resources and personnel from other locations as needed to help with emergency situations, and keep HQ and the local DMC in the loop about any changes.

Disaster Stage

(a) Assist with evacuation, rescue, and relief efforts while maintaining communication with local authority.
(b) Determine the extent of the damage, provide repair and rebuilding cost estimates, and get the necessary funding.
(c) Help save the lives of those who are in danger.
(d) Help in the relocation of stock and other assets if required.

Rehabilitation Stage

(a) Get started right away on fixing up government buildings that were destroyed in the calamity.
(b) Start planning for the future of government buildings that have been destroyed, and go to start on repairs and restoration right once.
(c) To aid with the relief and restoration efforts, coordinate with local government and authorities.
(d) Create comprehensive plans, schedules, and budgets for safeguarding public assets in the event of another natural catastrophe and submit them to the relevant authorities.
(e) The rebuilding effort should be supervised and technical aid provided if needed.

Urban Development Authority (RAJUK, CDA, KDA, RDA & NHA)

Risk Reduction

(a) Have in touch with the GSB to make sure you get the seismic risk maps you need to identify risky regions. We need to include earthquake microzonation maps into city planning at the neighborhood level.
(b) Invent methods and provide technical aid to have high-rises built in BNBC-stricken regions.
(c) Urban development design for seismic disaster mitigation should include the special measures (provisions for broad roads to facilitate the smooth functioning of the transportation participating in the rescue effort, arrangement for construction of emergency shelters in the open/free space in every Para).
(d) Survey the region and compile seismic risk maps to eventually include all the buildings and other structures in the area. Having a backup plan is important.
(e) Collaborate with the locals to prevent the development of buildings and other structures that might impede the smooth operation of vehicles. Keep in touch with law enforcement to make sure it gets fully implemented.
(f) Engineer planning groups and construction companies should be given training in urban planning and building retrofitting.
(g) Put into action the BNBC to develop any structures or mansions in strategic locations.
(h) Help sort buildings and other structures into safe and dangerous categories.
(i) Help out with temporary storage and repairs if needed (especially for hospitals, emergency shelters, and so on).

(j) Apply the BNBC and the earthquake-resistant building code throughout the reconstruction phase.

Emergency Response

Normal Times

(a) Make sure the civil engineers working on the building and the masons understand the BNBC and how to implement it properly by holding training sessions and doing awareness campaigns.
(b) Put in place a rigorous mechanism for checking the quality of public and private construction projects.

Alert and Warning Stage

(a) Create the Command Center to coordinate with regional offices and DMCs.
(b) Once warning signs are received, an immediate alert must be sent to all parties involved.
(c) Maintain a ready force of guards and supplies to safeguard and fix up public buildings.
(d) If required, move all government supplies to a secure location to guarantee their safety.
(e) Send resources and personnel from other locations as needed to help with emergency situations, and keep HQ and the local DMC in the loop about any changes.

Disaster Stage

(a) Assist with evacuation, rescue, and relief efforts while maintaining communication with local authority.
(b) Determine the extent of the damage, provide repair and rebuilding cost estimates, and get the necessary funding.
(c) Help save the lives of those who are in danger.
(d) Help in the relocation of stock and other assets if required.

Rehabilitation Stage

(a) Get started right away on fixing up government buildings that were destroyed in the calamity.
(b) Start planning for the future of government buildings that have been destroyed, and go to start on repairs and restoration right once.
(c) To aid with the relief and restoration efforts, coordinate with local government and authorities.
(d) Create comprehensive plans, schedules, and budgets for safeguarding public assets in the event of another natural catastrophe and submit them to the relevant authorities.
(e) The rebuilding effort should be supervised and technical aid provided if needed.

Ministry of Establishment

Risk Reduction

(a) All government agencies should be required to update their human resource development policies and procedures to include disaster management.
(b) Make it mandatory for all public academies and universities to include a disaster management module into their curricula, and create a disaster management training course that takes into account the unique needs of each public service cadre in the event of a crisis.
(c) Initiate an interdepartmental and cross-ministry communication mechanism aimed at reducing risks.
(d) Create a plan for the Ministry to deal with potential threats and emergencies.
(e) Create a method for rapid reaction in certain sectors in the event of an emergency.
(f) Create a Ministry emergency plan.

Emergency Response

Normal Time

(a) Keep a reserve of trained personnel who are equipped to deal with large-scale catastrophes for immediate deployment to the impacted regions.
(b) Create a program to reward and keep the hardworking, competent people who are bringing services to remote locations.

Disaster Stage

(a) The places hit most by the tragedy need to have the most capable individuals sent there to deal with the aftermath.

Rehabilitation Stage

(a) Send in the experts who can take care of the rebuilding work.

4.4.5 Ministry of Finance

Risk Reduction

(a) A senior official in the Ministry will operate as the Focal Point of guaranteeing mobilization of the essential financial resources for all the actions linked to the Government's complete Disaster Risk Reduction methods and initiatives.
(b) You should appoint a high-ranking official to function as the ministry's point person on disaster management.
(c) Find one person to serve as a point of contact for disaster management in each Ministry department.
(d) Develop financing policies and create a disaster risk reduction fund as per the directive of the NDMC.

(e) Mainstream disaster management ideas and techniques into national financial development processes and policies.
(f) Ensure sufficient cash allocation with reference to the request made by relevant Ministries, Departments, and organizations.
(g) Get in touch with relevant Ministries, Committees, and Bodies to coordinate and coordinate financial assistance for their planned actions relating to Risk Reduction projects.
(h) Make that the relevant Ministries/Departments/Committees have the funding they need to conduct an Impact and Risk Assessment in the event of a natural disaster.
(i) Develop and execute a Disaster Management Training Policy for Finance Ministry workers.
(j) Develop, approve, and implement approaches for integrating disaster management planning for resources mobilization at different levels within financial development planning procedures.
(k) Ensure participation in National Disaster Management Council (NDMC), Inter-Ministry Disaster Management Coordination Committee (IMDMCC), and National Disaster Management Advisory Committee (NDMAC) sessions.
(l) Send out the appropriate directives and instructions to everyone involved with the Finance Ministry's decision-making process about funding needs.
(m) Take charge of and keep track of all financial records that fall within the CFO's purview (Finance Division).

Emergency Response

Normal Times

(a) Make and keep safe financial provisions for sudden needs in catastrophe risk reduction.
(b) Fund training, capacity building, warning system development/learning exercises, simulated drills, and other similar initiatives across government ministries, departments, Upazilas, UPs, etc.
(c) Ensure close cooperation between the MoFDM and the NDMC.
(d) Make a Ministry emergency plan.

Disaster Stage

(a) Make sure money is sent to where it needs to go quickly so that relief efforts and early recovery can get underway.
(b) Collaborate closely with MoFDM to compile data on disaster losses, including the types and quantities of items needed for relief efforts.
(c) Make sure that overseas aid is delivered on time by handling things like releasing it from customs on schedule.

Rehabilitation Stage

(a) Make sure money is quickly allocated so that restoration and possible rebuilding efforts may get under way as soon as possible.

(b) Keep an eye on how well the defined rules and procedures are being followed for the financing of public infrastructure, facilities, and buildings in hazard effect regions.
(c) Help the Foreign Ministry, the Ministry of Foreign Affairs, and the Ministry of Foreign Disaster Management.
(d) Be sure any plans for rebuilding or financing proposals take into account findings from post-disaster impact risk assessments.

Finance Division

The Financial Group will also be responsible for the following, in addition to its regular obligations.

Risk Reduction

(a) Appoint one Ministry Liaison Officer to serve as the Department's main point of contact for disaster management.
(b) Make sure the requested funds are available so that relevant agencies may carry out their plans and programs to reduce risk.
(c) Create the criteria for determining funding needs for the Ministry of Finance and the accompanying instructions and guidelines.
(d) Help the Ministry of Finance create, approve, and implement methods for incorporating disaster management planning for resource mobilization at different levels into financial development planning procedures.

Emergency Response

Normal Times

(a) Preparation in the form of a revenue budget allocation for emergency management.
(b) Get the money together as the NDMC has decided.

Disaster Stage

(a) Make sure money gets to relief efforts quickly.
(b) Assist the Chief Accounts Officer in managing all accounts within their purview by providing guidance and support as needed.
(c) Always be prepared to quickly mobilize resources to rebuild damaged infrastructure, public facilities, and buildings.

Rehabilitation Stage

(a) Take part in government-wide loss, damage, and needs evaluations; compile a single, approved budget for all ministries' repair and rehabilitation efforts.

Economic Relations Division

In addition to normal functions, the Economic Relations Division will perform the following duties.

Risk Reduction

(b) Make one of the division's Liaison Officers the primary point of contact for disaster management.
(c) Get suggestions about how to reduce risks in certain industries from the relevant government agencies.
(d) Raise money from other sources to help in catastrophe preparedness.

Emergency Response

Normal Times

(a) In order to quickly mobilize external help following a catastrophe, it is important to build networks with multilateral and bilateral partners and to collect cash for the disaster risk reduction and climate change impact fund project from development partners.
(b) Always keep track of the many external organizations that have helped in the past when disaster strikes.
(c) Create an overarching policy framework for receiving aid from other countries in times of need.

Disaster Stage

(a) Monitor the catastrophe scenario using DM&RD and DMB, and keep track of loss/damage figures.
(b) Information on the kind and quantity of relief supplies should be kept available. Help in drafting a letter asking for an international rescue mission if it becomes necessary. The data should be consistent with that produced by the Foreign Ministry.

Rehabilitation Stage

Seek international aid in accordance with directives from the DM&RD, MoFDM, and MoFA.

Internal Resources Division (National Board of Revenue)

The Internal Resources Department will take on certain additional responsibilities in addition to its regular work.

Risk Reduction

(a) Build a policy infrastructure to incentivize reduction of risks in all economic sectors, whether via tax breaks or some other methods.
(b) Create a protocol for expedited relief supplies to get through customs.
(c) Create a protocol for the customs duty of disaster-prevention and emergency response tools.
(d) Get the NBR's staff trained in disaster management.

Response Management

Normal Times

(a) Make one of the division's Liaison Officers the primary point of contact for disaster management.
(b) To expedite the arrival of aid supplies, please instruct Customs to do so as soon as possible.

Disaster and Rehabilitation Stage

(a) Quickly release all humanitarian aid equipment and commodities from airports and harbors by following official instructions for clearance and tax exemption.

Ministry of Planning

Risk Reduction

(a) Choose a high-ranking member of the commission to serve as the Focal Point and initiate communication with the NDMC, MOFDMC, and IMDMCC.
(b) Make that the Ministry of Planning is represented in NDMC and IMDMCC gatherings.
(c) Directives should be issued to ensure that development projects are planned and evaluated with catastrophe risk reduction and prior project lessons in mind.
(d) Guarantee enough resources (both technical and financial) to back up its risk management and mitigation efforts throughout the government. Strategies.
(e) Prioritize mitigation measures based on formal hazard and risk assessments. These projects might include, but are not limited to, embankments, forestation, and jetties on coastal islands, coastal roadways, telecommunications, and shelter places.

Emergency Response

Normal Times

(a) Create a Ministry-wide emergency response mechanism.
(b) Make a backup plan to help you cope with unexpected events and emergencies brought on by both natural and man-made disasters.

Disaster Stage

(a) Spending on relief efforts in the worst-hit regions should be prioritized, with funds being distributed to the relevant ministries.

Rehabilitation Stage

(a) Prepare a budget for infrastructure maintenance and renewal.
(b) Make that funding proposals for repairs and rehabilitation include a hazard and risk assessments of the area after the impact, and that measures to mitigate risks are adopted.

Planning Commission

The Planning Commission plays a crucial role in the prioritization of funding for disaster prevention and recovery efforts. Funding initiatives to reduce the impact of disasters and help victims recover is really funding development programs that would otherwise be dormant. Keep this in mind when you craft your growth plans and initiatives.

Risk Reduction

(a) Develop a guiding policy:
(b) Implementing catastrophe risk reduction measures in the planning, implementation, and evaluation of development programmes
(c) All investment projects and programs must be thoroughly vetted to make sure they won't expose the economy to any unanticipated dangers.
(d) For DRR to be included into the development planning process and institutionalized,
(e) budget more heavily for initiatives that would not expose people to new dangers or risks.
(f) Create a robust mechanism for keeping tabs on disaster preparedness to make sure the government follows its own rules.

Emergency Response

Normal Times

(a) Projects such as embankments, afforestation, jetty at coastal islands, roads in coastal regions (essential for evacuation and relief work), telecommunications, reinforced earthen mounds, and building of shelter places should be given suitable priority in the event of a catastrophe.

Rehabilitation Stage

(a) Consider the full extent of the harm done to infrastructure before allocating funding for its repair or restoration.

Ministry of Women and Children Affairs (MWCA)

Risk Reduction

(a) Appoint a high-ranking official to serve as the Ministry's main point of contact for disaster management.
(b) Start communicating often with the IMDMCC, the Ministry of Foreign Disaster Management, and any other relevant disaster management committees.
(c) Organize a capacity-building programme for workers at the Upazila and UP level, with the participation of workers/officials from the Ministry of Social Welfare, the Ministry of Local Government and Rural Development, and the Department of Management and Budget (DMB), to reduce risks to women and children and to ensure their protection rights.

(d) Create a strategy for reducing risks and increasing preparation across all industries.
(e) All MWCA development plans and programs should include measures to reduce and manage risk.
(f) Put together some kind of checklist or indicators to make sure that women's and children's (and especially teenagers') empowerment is a part of all community development plans and programs.
(g) Prepare for disasters by doing a risk assessment and implementing a risk reduction program, ensuring that relief supplies, as well as protection and psychological support packages for children and adolescents, are properly prepositioned.
(h) Make sure there is money in the budget for a short-term response program that focuses on reducing women and children's vulnerability right away.

Emergency Response

Normal Times

(a) Take special care to ensure that the newly built cyclone shelters provide separate facilities for women, children, the old, and the handicapped.
(b) Be sure that the Ministry's various awareness-raising training programs include drills and exercises for dealing with disasters.
(c) All proposed methods of producing revenue should be designed with risk avoidance and contingency planning in mind.
(d) Create a system of community-based emergency services.
(e) Organize a Ministry backup plan.

Warning/Alert Stage

(a) Collaborate with BMD and MoFDM to get the word out in a timely manner so that vulnerable populations (such as women, children, the elderly, and the crippled) may safely evacuate.

Disaster Stage

(b) Work with the Local Government Division, Department of Social Services, DMB, and Shishu Academy to provide resources (both financial and human) to conduct an immediate evaluation of the impacted population's vulnerability, with a focus on women and children.
(c) Assist in coordinating and regularly reviewing services for women, children, and other marginalized populations.
(d) Make it a priority to set up a system for collecting data and sharing it on the most pressing protection issues, aid gaps, and ways to fill them.
(e) Raise public awareness on the need of protecting children and women, and of making sure that their rights are being upheld.

Rehabilitation

(a) Make that the rehabilitation plans of all government agencies take into account the special needs and priorities of women, children, and the elderly.

Department of Women Affairs

Besides the normal function, the department will carry out the task on disaster management.

(b) Make sure there's departmental representation on all the disaster-related committees.
(c) Make sure women are included in emergency planning and response.
(d) Do everything you can to close the gender gap in disaster management and make sure it's implemented in all your plans.
(e) Help mothers and children who have lost their homes recover economically.

Ministry of Law, Justice, and Parliamentary Affairs

Risk Reduction

(a) Appoint a high-ranking member of staff to serve as the Ministry's point person on disaster risk management issues.
(b) Legal counsel for the MoFDM and DMB.
(c) Improve the time it takes for the NDMC to approve the National Disaster Risk Management Policy Framework by coordinating closely with the PM office.
(d) Help the PM office speed up the process of approving or endorsing the Disaster Management Act, the SOD, and any other legal papers dealing with disaster risk reduction and emergency response.
(e) Determine the risks facing each industry, and work with the Ministry to create a strategy for mitigating those risks.
(f) Protect the Ministry's ability to allocate resources toward risk mitigation and management in the annual budget.

Emergency Response

Normal Times

(a) Create a policy framework for handling any inheritance, property, or guardianship concerns that may emerge after a catastrophe.
(b) Set up a method of sharing information about potential dangers across industries.
(c) Build the Ministry's strategy for preventing and responding to potential disasters.
(d) Create a Ministry-wide emergency response mechanism for certain industries.

Disaster Stage

(a) Assist in the creation of laws and make sure the Ministry is adequately represented at the IMDMC and NDMC meetings.
(b) Keeping parliament informed of developments and ensuring appropriate support.

Ministry of Social Welfare

In addition to carrying out the Ministry's own work plans and performing routine administrative tasks, the Secretary of the Ministry of Social Welfare will also be responsible for the following.

Risk Reduction

(a) Appoint one staff member as the Ministry's primary contact for disaster management.
(b) Create a strategy for implementing the agreed, accepted, and implemented action plan for risk reduction and disaster preparation for the vulnerable population.
(c) Educate and train Upazila social services officials and union social workers on child protection issues such as early identification, case management, psychological assistance, and safeguarding against abuse, violence, and exploitation.
(d) Those volunteer groups who are recruited and educated on disaster management should also be educated on how to safeguard children in times of crisis.
(e) Obtaining and securely prepositioning essential goods, such as family kits, education and psychological support kits, with key locations and necessary organizations.

Emergency Response

Normal Times

(a) Assist in creating, reviewing, and disseminating to the relevant administrative bodies a yearly disaster preparation and response plan that takes into account the unique needs of women and children.
(b) Appoint one Ministry Liaison Officer to serve as the Department's main point of contact for disaster management.
(c) Make use of the Directorate of Social Service to organize volunteer groups, then instruct them in disaster management programs and ensure they always have the supplies and equipment they'll need on hand.
(d) Coordinate personal work plans with all tiers of administrative authority and conduct yearly reviews of catastrophe preparation plans with the relevant Social Welfare office.
(e) Families without land who are forced to live close to embankments should benefit from new and improved programs that provide them with vocational training.
(f) In the event of a catastrophic event, social service providers should be organized by geographic region.

Alert and Warning Stage

(a) Maintain constant (24/7) communication with all parties involved, and keep the MoFDM abreast of major developments.
(b) Establish one Liaison Officer whose job it is to maintain communication with the EOC at the MoFDM and the DMB.

(c) As directed, help the local government via its own field officers and agencies in alerting people to disaster readiness, with special attention paid to the needs of children, women, and the crippled and old.

Disaster Stage

(a) Social workers should be called in via the Department of Social Services to help care for impacted communities, and emergency aid should be provided, including structural care and psychological support for children.
(b) Coordinate with local government, non-governmental organizations, and volunteer groups to set up and run relief centers and child-friendly spaces (CFS).
(c) Provide medical and other assistance to orphans in impacted regions, as well as safeguard them and help them adjust to life in the community or other care facilities based on what is in their best interest.
(d) Tend the MoFDM's Control Room and maintain contact with the DMB and EOC.
(e) Gather employees and volunteers to do an assessment of the loss and damage, keeping in mind the special needs of women, children, and the disabled.
(f) Child-friendly settings and referral services must be coordinated, managed, and monitored by social workers who are sent to impacted regions as needed.

Rehabilitation Stage

(a) Help safeguard vulnerable populations from damage and ensure they have fair access to essential services and livelihood assistance as part of a coordinated rescue, relief, and rehabilitation effort.
(b) Improve the ability of relevant line departments and NGOs to deal with problems of recovery and rehabilitation in a way that is sensitive to the needs of children and women.
(c) To the greatest extent possible, orphaned and separated children should be allowed to stay with and be cared for by their families/caregivers, and measures should be taken to introduce and strengthen family and community-based approaches to the protection, rehabilitation, and reintegration of vulnerable individuals. Take your child into an institution or orphanage as a last option and for as little time as feasible.
(d) To stop children from being separated from their families throughout the repatriation and resettlement process, it is important to put up local/community-based procedures for monitoring, reporting, and following up with vulnerable persons.
(e) Create a workable strategy for the economic and social reintegration of the crippled, widows, and their children, and present it to the appropriate authorities.
(f) Help get the government's and other authorities' reconstruction plans off the ground by lending them your technical expertise, and keep going until things are back to normal.

Department of Social Services

Besides the normal function, the department will carry out these tasks on disaster management:

(a) Make sure that the department is well-represented on various disaster management boards.
(b) Make that people with disabilities and other marginalized groups are included in emergency planning and response.
(c) Contribute to the search-and-rescue operation and the safe evacuation of handicapped people.
(d) Send out an order to the field office to make sure that the search-and-rescue operation is handled with care for those with special needs.
(e) Help those in need in the wake of natural disasters by funding various programs that aid the crippled and underprivileged.
(f) Send updates on the department's success in coordinating disaster risk reduction and emergency response efforts to the DMB.

Ministry of Shipping

When a disaster strikes, the Ministry of Shipping must go above and beyond its normal duties to ensure that the water transports of the Shipping Directorate, the Bangladesh Inland Water Transport Authority, the Bangladesh Inland Water Transport Corporation, and private owners of disaster-affected areas are assembled and put into service as soon as possible to meet the needs of the Ministry of Food and Disaster Management and other government agencies.

Risk Reduction

(a) Pick a high-ranking official to act as the Ministry's main point of contact for disaster management.
(b) Create a strategy for dealing with potential threats and being ready for them.
(c) Prepare for and react to disasters by building and maintaining jettys, dredging and cleaning rivers, and organizing relief efforts (maintain list of suitable shipping).
(d) Make sure the Ministry has enough money to fund its disaster management efforts.
(e) Conduct a Ministry-wide risk analysis.
(f) Create a strategy to lower risks in your industry.

Emergency Response

Normal Times

(a) Please appoint one Liaison Officer to serve as the Ministry's primary point of contact for disaster management.
(b) Keep track of all the ships owned by BIWTA, BWTC, and individuals. The list will include the names and addresses of the owners, allowing for urgent requests to be fulfilled. These ships and vessels should be principally used for the following duties:
    (i) Evacuation of people prior to disaster
    (ii) Evacuation of marooned and distressed people and livestock

(iii) Transportation of food grains from food godowns
(iv) Transportation of relief materials, medical supplies, and people engaged in relief work
(v) Maintain communication systems and re-establish the same where it has been snapped

(c) The rescue and relief ships need access to the islands, thus it is necessary to build jetties and keep the waterways clear.
(d) Maintaining open waterways by routinely dredging them and clearing any obstructions would allow ships and boats to more freely carry out evacuation, rescue, and relief operations in times of crisis.

Alert and Warning Stage

(a) After receiving the alert, set up the Control Room and hire the necessary personnel.
(b) Choose one officer to serve as the EOC's main point of contact at the MoFDM's Liaison Office.
(c) The connection to BIWTA, BIWTC, and all ports must be kept open.
(d) If a warning signal is received in accordance with the prediction, the closest stations should be identified and the ships should be gathered from the potentially impacted regions and relocated to a safer location.
(e) If necessary, make plans to request private water transportation from the relevant authorities.
(f) Send out a warning to all outposts and departments below you so that they can get ready.
(g) Help the authorities with the logistics of relocating people and animals to secure areas.
(h) Assist in any way you can with evacuations, rescues, and relief efforts by coordinating with local authorities and other ministries.
(i) Maintain repair crews in port and on fast-moving watercraft with all necessary tools and supplies.
(j) Maintain a secure system for transporting ships.
(k) Maintain a well-oiled ferry service.
(l) Coasters should be repaired and used for transporting aid and emergency supplies.
(m) Direct the Port Authority to give precedence to the arrival of boats carrying food and other necessities, as well as the delivery of their cargo.
(n) If ships are destroyed in a catastrophe, make sure they can be repaired quickly and that POL is readily available.
(o) Protect your own facilities and move any mobile equipment, supplies, and other installations to a more secure location.
(p) Daily updates on information obtained via own channels about the cyclone status and all actions should be sent to the EOC at the MoFDM.

(q) Maintain waterborne transportation at designated outposts outside of dangerous zones, and relocate as needed. Coordinate ahead of time with relevant Ministries and local authorities.

Disaster Stage

(a) Maintain Control Room operations seven days a week, twenty-four hours a day, in the event of an emergency.
(b) Make ships available in disaster zones and use them as needed by local authorities, the MoFDM, and other government organizations.
(c) If necessary, ships might be sent from elsewhere to the afflicted locations.
(d) Act now to evacuate yourself and your own facilities, supplies, and equipment from the threatened locations.
(e) Instruct the relevant port authorities to ensure the prompt delivery of donated or imported food and other necessities.
(f) Ascertain the extent of the damage to your own property, compile cost estimates, and make plans for its restoration or replacement.
(g) Assist disaster-stricken vessels in getting back on the water as soon as possible by facilitating their access to POL.

Rehabilitation Stage

(a) Repair the water supply and distribution in the region that was damaged.
(b) Keep up the good work with the relief and recovery initiatives.
(c) Prepare the vessel owned by the BIWTC and the boats chartered by the chartering committee to carry the relief and food supplies received from foreign nations under import or grant and to transfer the required amount of goods inside the country.
(d) Marine craft should be stationed in unaffected locations in case they need to be sent to disaster zones. Consult with relevant local departments and other Ministries on this matter.

Bangladesh Inland Water Transport Corporation (BIWTC)

Besides normal work of the corporation, BIWTC will perform the following duties.

Risk Reduction

(a) Make sure that all the boats that will be utilized for evacuation, transporting relief materials, and managing relief activities are recorded in a database.
(b) Provided funding in the budget for maritime infrastructure upkeep and repair.
(c) Perform a threat analysis and locate all of the jetties, ships, and ferries that might be in danger.
(d) Maintain funding for the Ministry's risk reduction and preparation initiatives.
(e) Develop training and informational initiatives to inform Ministry and agency personnel about disaster preparedness measures.

Emergency Response

Normal Times

(a) Assign a Liaison Officer to serve as the company's primary point of contact for all matters relating to disaster management.
(b) Prepare for any potential calamity by making all water transportation and evacuation facilities available to the authorities.
(c) Those who are stranded or whose livestock is in danger should have access to water shipments on demand.
(d) Please provide water transfers upon request for the following uses:

(i) transporting food grains to storage facilities and then shipping them out
(ii) to relocate emergency supplies, medical personnel, and relief workers
(iii) keep communication going and fix the broken connection

(e) Assist with the rapid transport of relief supplies by ferry if needed.
(f) Help in the acquisition of personal watercraft.

Alert and Warning Stage

(a) Set up a command center at the BIWTC's main office and keep in touch with regional departments through their own private network.
(b) Appoint one officer to serve as the liaison between the Ministry of Transportation, the Ministry of Foreign Affairs and Development, and the Department of Maritime Affairs.
(c) Following the receipt of a cyclone warning, water transports from the concerned region should congregate at a single, readily accessible port located as close as possible to the area expected to be impacted by the catastrophe.
(d) Create a communication connection with all ports, send out an alarm to all subordinate offices, take precautions to secure your own installations, stores, and equipment, and relocate comparable transportable installations, stores, and equipment to a safer location.
(e) Make sure you always have a repair staff on standby, well-stocked with the ability to act quickly, and all the necessary supplies in water vessels.
(f) Ships should be ready to evacuate passengers to the closest safe areas.
(g) Help the government organize and carry out evacuation, rescue, and relief efforts in coordination with local authorities.
(h) Keep all ships and ferries operating in a risk-free manner.
(i) Get the necessary number of BIWTC Coasters repaired so that they may be used to convey aid and emergency rations in times of crisis.
(j) Send daily updates to the EOC at the Ministry of Food and Disaster Management, with a copy to the Ministry of Shipping, detailing the cyclone status and any additional activity learned of from your own sources.
(k) In coordination with the Ministry of Shipping, the Ministry of Foreign Affairs and Development, and local government officials, keep ships and water transport on standby at prefixed stations in unaffected areas, and move them to convenient places for evacuation, rescue, and relief operations if necessary.

Disaster Stage

(a) Maintain the Control Room's regular, round-the-clock schedule.
(b) Maintain the integrity of the ships operating in the affected regions, and put them to use in accordance with the needs of the local civic authority, the Ministry of Food and Disaster Management, and other government agencies.
(c) The impacted locations should have ships sent there from other stations if required.
(d) Steps should be taken to ensure the safety of one's own facilities, stocks, and machinery in the impacted locations.
(e) Plan for swift ship repairs and simple POL restocking.

Rehabilitation Stage

(a) All available resources should be allocated to relief and restoration efforts.
(b) Ships owned or chartered by the BIWTC should be directed to convey aid and food supplies.
(c) Prepare for a full catastrophe assessment of BIWTC facilities, machinery, and vessels, and then take all necessary measures to recover losses and resume operations as soon as possible.

Bangladesh Inland Water Transport Authority (BIWTA)

In addition to normal activities, BIWTA will perform the following duties.

Risk Reduction

(a) Determine which BIWTA Liaison Officer will serve as the primary point of contact for all matters pertaining to disaster management.
(b) Assist with project planning for rescue, relief, and rehabilitation efforts, and participate in Disaster Management Committee meetings.
(c) Protect your ships, shore facilities, signals, waterway markings, lighthouse, and buoys by taking precautions.

Emergency Response

Normal Times

(a) Set up and maintain waterway markers, signals, buoys, and lighthouses, and dredge the shore on a regular basis to enhance maritime communication.
(b) Increase the size of the maritime rescue fleet and ensure it is always well stocked and ready to go.

Alert and Warning Stage

(a) Establish a single Control Room at the BIWTA Main Office and staff it with competent individuals.
(b) Find a single contact person at the MoFDM, the MoS, and the BIWTC to serve as the EOC's main point of contact.

(c) Tide Guess Monitoring Boaya should share the data it has gathered with BMD, DMB, and FFWC.
(d) In order to have everything ready, notify the relevant subordinate offices and installations.
(e) Prepare a rescue crew with all the supplies they'll need and keep them on standby.
(f) Protect your own facilities, supplies, equipment, etc., by taking the required precautions, including relocating any mobile components to a more secure location.
(g) Regular updates on damage/loss and rebuilding should be sent to IMDMCC and MoFDM with a copy to the Ministry of Shipping.

Disaster Stage

(a) Maintain Control Room operations seven days a week, 24 h a day.
(b) Evacuation, rescue, and relief operations in an emergency call for close coordination with local authorities, and the provision of whatever aid required to ensure safe waterways for ships and water transportation.
(c) Prepare a rescue fleet with the necessary gear and have them stationed at secure locations close to the disaster zone.

Rehabilitation Stage

(a) Estimate the cost of rebuilding or repairing your ship's jetty, stations, installations, etc.
(b) Regular reports on losses, damages, and repairs should be sent to IMDMCC and MoFDM, with a copy to the Ministry of Shipping.
(c) Plan and estimate the costs of short- and long-term maintenance and replacement of directional indicators like buoys and lighthouses.
(d) Rapid transfer of aid supplies requires immediate action to recover sunken boats and restore dislodged coastal line signs.
(e) After all rescue and rebuilding efforts have been made, please provide a full report to the NDMC/MoFDM and the Ministry of Shipping.

Department of Shipping

In addition to normal activities, Department of Shipping will perform the following duties.

(a) It is important to designate one Liaison Officer within the Department to serve as the focal point for disaster management concerns.
(b) Take part in the Disaster Management Committee meeting.
(c) Keep an eye on the international shipping of potentially dangerous goods.
(d) Safeguards against the release of potentially polluting substances like oil spills, radioactive sludge, or harmful organisms into the ocean must be in place.
(e) Keep in touch with relevant international maritime groups to coordinate the gathering and sharing of data on potential maritime risks.
(f) Make a backup plan in case of major emergencies.

Chittagong Port Authority

In addition to normal activities, the port authorities will perform the following duties:

(a) Select one Liaison Officer to serve as the authority's main point of contact for all disaster-related matters under Disaster Management.
(b) Maintain the security of berthing and cargo terminals even in the event of a major emergency.
(c) Make sure the port has a backup plan in case something goes wrong.
(d) Make sure that the port's infrastructure (cargo handling, warehouses) is ready to assist disaster response operations as quickly as possible following a catastrophe.
(e) Help provide essential supplies and tools to disaster zones quickly so that people may start searching for survivors and rescuing those who have been trapped.
(f) After receiving early warning and signals, immediately notify all involved parties and ships.

Mongla Port Authority

In addition to normal activities, the port authorities will perform the following duties.

(a) Select one Liaison Officer to serve as the authority's main point of contact for all disaster-related matters under Disaster Management.
(b) Maintain the security of berthing and cargo terminals even in the event of a major emergency.
(c) Make sure the port has a backup plan in case something goes wrong.
(d) Make sure that the port's infrastructure (cargo handling, warehouses) is ready to assist disaster response operations as quickly as possible following a catastrophe.
(e) Help provide supplies and tools for rescue missions to those who need them quickly after a natural catastrophe.
(f) After receiving timely signals and warnings, immediately notify all involved parties and vessels.

Ministry of Communications

The Ministry of Communications is responsible for ensuring that all lines of communication are always open, and in the event of a catastrophe, this includes keeping the rail lines open to the closest train lines so that supplies can be sent to the affected regions. Following are the responsibilities of the Ministry and its divisions as stipulated by the SOD.

Risk Reduction

(a) Pick one high-ranking official to act as the Ministry's "Focal Point for Disaster Management".
(b) Instruct all departments to collaborate on creating a catastrophe risk management communication guideline.
(c) Connect closely with the MoFDM and the NDMC.

(d) Build the Ministry's strategy for preventing and responding to potential disasters.
(e) Make sure a strategy to reduce risks in a certain industry is developed and put into effect.
(f) Put aside enough money in the budget for infrastructure upkeep and repairs.
(g) Make sure that BWDB and the rest of the Ministry are on the same page with their risk mitigation strategies.
(h) Raising railway lines, roadways, fortifying culverts, and reinforcing bridges are all examples of risk reduction actions that may be carried out to ensure that infrastructure is protected from damage.
(i) Create a method for the Ministry to share information about potential dangers.
(j) Create a method for rapid reaction in certain sectors in the event of an emergency.
(k) Make sure your business has a backup plan in case of any emergency.

Emergency Response

Normal Times

(a) Plan and direct the use of transportation and communication networks to facilitate emergency response, disaster relief, and reconstruction efforts, giving priority to the conveyance of relief supplies and personnel as needed.

Alert and Warning Stage

(a) Make sure the warning is getting out there.

Disaster Stage

(a) Keep the lines of communication open in the regions hit by the catastrophe so that immediate measures may be taken to evaluate the extent of the damage and begin work on a quality rebuilding.
(b) Facilitate the deployment of extra personnel and supplies to disaster zones, including the deployment of members of the armed forces to perform communication roles.
(c) Maintain all services and infrastructure, and if anything goes wrong, make sure you have a backup plan for getting things done.
(d) Build makeshift bridges and connecting roadways where they are needed.
(e) Keep an eye out for any signs of damage in critical infrastructure and report any issues as soon as possible.
(f) If required, it should reorganize its resources, including its personnel.

Rehabilitation Stage

(a) Prepare for the worst and start fixing the infrastructure as soon as possible.
(b) Maintain open lines of communication, accurate reporting, and timely information throughout disaster response and recovery.
(c) Protect vital services and infrastructure by whatever means necessary.

Bangladesh Bridge Authority

The Ministry of Communications is responsible for ensuring that all lines of communication are always open, and in the event of a catastrophe, this includes keeping the rail lines open to the closest train lines so that supplies can be sent to the affected regions. With respect to the SOD, the Ministry of Transport's Roads and Railways Division and Jamuna Bridge Division will be responsible for the following.

(a) Get together a backup plan for the organization.
(b) Do not stop using the agency's backup plan.
(c) Involve the Geological Survey of Bangladesh and the Bangladesh Meteorological Department in analyzing data from seismic monitoring equipment and Galmitre.
(d) After major earthquakes or other catastrophes, communicate your observations to the appropriate authorities and make sure that BBA infrastructures are being monitored on a consistent basis.

Bangladesh Railway

Risk Reduction

(a) Determine which Liaison Officer will serve as the organization's Focal Point and share that information with everyone involved.
(b) Provide all railway departments with a set of rules to follow when dealing with a catastrophe.
(c) Any new infrastructure has to take into account both potential future dangers and existing ones.
(d) In the event of a natural catastrophe, the government will expect agencies to have a backup plan ready to implement as soon as possible.
(e) Make sure that the most important vehicles and equipment for disaster recovery are used for SAR and relief efforts.
(f) Do not stop using the agency's backup plan.
(g) Involve the Geological Survey of Bangladesh and the Bangladesh Meteorological Department in analyzing data from seismic monitoring equipment and Galmitre.

Emergency Response

Normal Times

Review the current emergency plans in light of past disasters.

(a) Collaborate with the Ministry of Energy and Mineral Resources, the Ministry of Water Transport, the Jamuna Bridge Division, and the Roads and Railways Division to ensure the safety of all parties.
(b) Improve railway infrastructure by repairing tracks, elevating embankments, and reinforcing culverts and bridges.

(c) In the event of an emergency, provide orders to the relevant railway authorities to ensure that patrol guards are stationed at the railway's weakest spots.

Alert and Warning Stage

(a) The Railroad Headquarters should have a Control Room.
(b) Send out warnings to the train networks serving disaster-prone locations.
(c) Appoint a Liaison Officer to serve as a Point of Contact Between the EOC at the MoFDM and the rest of the Organization.
(d) When a tragedy occurs on a railway, the appropriate officer will use the railway's emergency communication network to alert the proper authorities.
(e) In the event of a catastrophe, it is important to have a backup plan for the amended schedule of train services.
(f) Fixing railways and bridges requires a stock of building supplies.

Disaster Stage

(a) On the basis of information gathered through railway channels, notify the NDMC, the IMDMCC, and the MoFDM.
(b) Improve railway safety by increasing patrols and inspecting bridges and tracks more often.
(c) Provide assistance and guidance to the relevant authorities in maintaining traffic and shifting it when necessary to ensure safety.
(d) Prevent the loss or destruction of any stored goods, machinery, tools, train stock, or locomotives by taking the necessary precautions.
(e) Make plans for those impacted to be transported to nearby train terminals.

Rehabilitation Stage

(a) Collect data about railway properties.
(b) Help the appropriate authorities determine the scope of the damage and the needs of the impacted population.
(c) For the benefit of NDMC and IMDMCC, please provide a list of total losses and damages to the EOC at the MoFDM.
(d) As soon as possible, fix the broken rails and resume regular service.
(e) Help the government in the area with their rescue, aid, and recovery efforts.
(f) If major junction stations or equipment are damaged, you should be prepared to restore them to working order as soon as possible.
(g) Prioritize the shipment of food and other humanitarian supplies as needed.

Bangladesh Road Transport Authority

In the aftermath of a natural catastrophe, it is crucial to ensure that relief personnel and supplies can go where they need to go without interruption. According to the SOD, the Bangladesh Road Transport Authority is responsible for ensuring this.

Risk Reduction

(a) Assign a single office Liaison Officer the role of primary contact.

(b) Any new infrastructure has to take into account both potential future dangers and existing ones.
(c) Get together a backup plan for the organization.
(d) Do not stop using the agency's backup plan.
(e) Distribute to GSB and BMD the agency readings obtained from Galmitre and other seismic monitoring equipment.

Emergency Response

Normal Times

(a) Protecting vital roadway transportation requires the issuing of proper rules for all parties involved within the authority.
(b) Bridges, culverts, roadways, and embankments must be regularly maintained and repaired with enough funding to resist the force of tidal bore or flooding after powerful storms.
(c) Evaluate current safety precautions in the context of historical events.

Disaster Stage

(a) Install a permanent Control Room (24 h).
(b) Keep in constant contact with the Emergency Operations Center at the Ministry of Agriculture and Disaster Relief.
(c) Forwarded to IMDMCC and MoFDM through own channel: very vital information.
(d) Increasing the number of patrols on strategic bridges and roadways is a must.
(e) If there is a breakdown in road communication, make the necessary repairs and continue transportation through other routes.
(f) Assemble and expeditiously dispatch all supplies for mending damaged infrastructure including roads and bridges.
(g) If you think it is essential, plan any other event you want.

Rehabilitation Stage

(a) If the road you're on becomes impassable, you must promptly make preparations for an alternate path.
(b) Roads and bridges need quick attention to begin the process of repair and rehabilitation.
(c) Assign higher priority to vehicles transporting aid personnel and supplies.
(d) Evaluate the state of the roads, bridges, and culverts, and submit your findings to the NDMC and MoFDM, while also requesting more funding if you think it will be required to deal with the emergency.
(e) Put up your best effort to restore communication, both temporarily and permanently, for the sake of the economic recovery of the cyclone- and flood-stricken regions and the well-being of the people who live there.

Roads and Highways Department

The following responsibilities, in addition to those normally assigned to the Chief Engineer of the Roads and Highways Department and the offices under him (as per delegation of authorities), will fall on his shoulders.

Risk Reduction

(a) When planning any infrastructure, past and potential disasters should be taken into account.
(b) To prevent damage from tidal bore/high floods caused by storms, it is necessary to reinforce roadways, embankments, light bridges, and culverts.
(c) Get together a backup plan for the organization.
(d) Do not stop using the agency's backup plan.
(e) Involve the Geological Survey of Bangladesh and the Bangladesh Meteorological Department in analyzing data from seismic monitoring equipment and Galmitre.

Emergency Response

Normal Times

(a) Ensure that the Roads and Highways Directorate's maintenance program for critical road communications continues without interruption by issuing directives to authorities at all levels.
(b) In disaster-stricken communities, it is essential to fix the roads that have deteriorated the most and keep them in good shape.
(c) Twice a year, reflect on the lessons learned from the previous year's emergency preparations and make adjustments as necessary.
(d) If required, send orders to build connecting roadways and makeshift bridges, and launch ferry services.
(e) Prepare for a storm, tide bore, or flood by assigning patrol duty to vital embankments, bridges, and roadways.
(f) You should always have your road-building tools on hand in case of an unexpected repair.

Alert and Warning Stage

(a) Create a single Liaison Officer position at the MoFDM to serve as the point of contact with the EOC.
(b) Assemble transportation and maintenance equipment, if needed, and dispatch them to strategic facilities.
(c) Prevent the waste, damage, and destruction of infrastructure, building materials, equipment, shops, etc.
(d) Having backup routes planned is important.

Disaster Stage

(a) Maintain constant Control Room operation (24 h).
(b) Keep in constant contact with the Emergency Operations Center at the Ministry of Agriculture and Disaster Relief.
(c) The National Disaster Management Council is responsible for informing the Disaster Management Bureau of any relevant information it receives.
(d) Increase the number of people on patrol around major highways and bridges.
(e) In the case of a road communication interruption, make plans for other routes of travel.
(f) Move swiftly to ensure the safety of all ferry infrastructure, supplies, and personnel in the affected region.
(g) Do anything else is required.

Rehabilitation Stage

(a) Roads and bridges that have been damaged or destroyed must be repaired right away.
(b) Build new pathways so that the communication network may function normally again.
(c) Transportation of aid personnel and supplies should take precedence over other loads.
(d) Quantify losses and submit reports to the NDMC and MoFDM, where you may also request more money if you need it to deal with the issue.
(e) For the sake of the affected population and the region's economic growth, all possible measures should be made and maintained to keep road communications operational, both temporarily and permanently.

Bangladesh Road Transport Corporation

Bangladesh Road Transport Corporation will perform the following disaster-related activities in addition to normal duties.

(a) In locations served by roads, evacuation transportation must be organized in accordance with government guidelines.
(b) Help the local government in any way possible with their efforts to rescue, evacuate, provide aid, and restore the area.
(c) On government direction, provide a fleet of trucks to convey aid supplies, rebuilding materials, and debris removal equipment to the impacted districts.

Ministry of Industries

In addition to its normal functions, the Ministry of Industries will perform the following duties.

Risk Reduction

(a) Choose a capable senior worker to serve as the Ministry's Point Person.
(b) Guarantee attendance at both IMDMCC and DMC gatherings.

(c) Analyze the effects of risks at the regional and national levels of industry using a rigorous hazard and risk assessment.
(d) Integrate the Ministry's development planning procedures with disaster management concepts and best practices.
(e) Create study plans for the Ministry of Hazardous Substances and Health and Environment.
(f) Always keep the NDMC updated on the status of sectoral risk reduction actions and programs.
(g) Incorporate concerns about catastrophe risk management into the planning stages of any new industrial or manufacturing facilities.
(h) Develop business continuity policies and processes at industry and higher levels.
(i) Develop an industry-specific strategy for mitigating risk.
(j) Make sure the Ministry has enough money set out for disaster management.
(k) Put in place a Ministry of Risk Management and Communication risk communication system.
(l) Create a method for rapid reaction in certain sectors in the event of an emergency.

Emergency Response

Normal Times

(a) Make sure that, well in advance of cyclone and flood seasons, high-risk area companies and industries engage in preparation activities such awareness and education programs, reviews of emergency plans at the factory/establishment level, and exercise and testing of plans.
(b) Make it a point that any new factories or businesses located in high-risk locations do thorough hazard and risk assessments and create robust business continuity strategies in response.
(c) Appoint one Ministry Liaison Officer to serve as the Department's main point of contact for disaster management.
(d) Be sure that you've taken every precaution to protect your workforce, tools, inventory, installation, factory, etc.
(e) Industrial concerns in cyclone- and flood-prone locations need the implementation of structure-based security measures.
(f) Workers in all businesses and institutions located in flood- and cyclone-prone regions should undergo mandatory training on disaster preparation programs to be implemented at various levels.
(g) Make sure your business has a backup plan in case of any emergency.

Disaster Stage

(a) Tell everyone to help out the local government as much as they can with rescue, evacuation, relief, and restoration efforts.
(b) Companies, governments, and business owners with facilities in cyclone- and flood-prone locations should practice protecting and maintaining their facilities in the event of a catastrophe.

Rehabilitation Stage

(a) Conduct a post-disaster damage assessment to ascertain the extent of the destruction and the amount needed to restore the impacted industries.
(b) Make that industry repair and rehabilitation financing bids contain risk treatment solutions and a post-impact hazard and risk analysis is conducted.
(c) Estimate the costs that may be incurred to restore damaged industrial operations.
(d) Get the injured businesses to submit project plans to the Planning Commission/Ministry so that they may get the money they need to recover.
(e) Help impacted businesses approach the Ministry of Planning and the Planning Commission for financing to address post-disaster repair and rehabilitation needs,
(f) Control and distribute money and other resources.

Ministry of Education

In addition to carrying out its regular responsibilities and implementing its own Action Plans (contingency plans), this Ministry will also be responsible for the following.

Risk Reduction

(a) Pick a high-ranking official to act as the Ministry's main point of contact for disaster management.
(b) Include disaster-related themes in the curriculum of all K-12, collegiate, post-secondary, and graduate institutions in conjunction with the DMB, DM&RD.
(c) The construction of new schools and other educational facilities should take into account hazard and risk maps to protect against potential threats in the future, such as earthquakes.
(d) It is imperative that any school buildings utilized as emergency shelters have separate facilities for men and women.
(e) In cyclone- and flood-prone locations, schools should be built to BNBC standards with at least two stories.
(f) Create a sector-specific strategy for mitigating risks and managing them.
(g) Facilitate the Ministry's evaluation of risks in various sectors.
(h) Make that the Ministry's efforts to mitigate risk are adequately funded.
(i) At least twice a year, all schools should do catastrophe safety, evacuation, and first aid exercises.

Emergency Response

Normal Times

(a) Assign a single Ministry Liaison Officer the role of "Focal Point" in Disaster Management.
(b) Connect with the International Medical Device Manufacturers and Congress and the National Defense Medical Association.

(c) Conduct reaction and recovery exercises in collaboration with CPP every April and September, and raise awareness and educate instructors and students in high-risk locations on what to do during response and recovery operations.
(d) Inspire and mobilize teachers and students in affected regions to participate in rescue, evacuation, and relief efforts by providing them with predisaster training on what to do at various points in the disaster's progression.
(e) Schools in cyclone- and flood-prone locations should be built with at least two stories wherever feasible.
(f) Make sure that elementary school buildings in cyclone-prone locations are properly repaired and maintained.
(g) Education institutions participating in the CPP should conduct large-scale drills for disaster preparation in April and September of each year to raise awareness and knowledge among the local population.
(h) Together with the DMB, incorporate disaster-related topics into K-12 and higher education courses.
(i) Have the Ministry's disaster management preparedness plans ready.

Disaster Stage

(a) Put the requisite number of schools, together with their buildings, under the jurisdiction of the local government so that they may be used as a refuge and relief center in the case of a hurricane or flood.
(b) Coordinate with local authorities to recruit educators and students as rescue, evacuation, and relief volunteers.

Rehabilitation Stage

(a) Assess the damage to educational facilities and provide restoration suggestions.

Ministry of Primary and Mass Education

In addition to carrying out its regular responsibilities and implementing its own Action Plans (contingency plans), this Ministry will also be responsible for the following.

Risk Reduction

(a) Appoint a single competent official within the Ministry to act as the Department's primary contact for disaster management.
(b) One person should be appointed as the division's liaison officer.
(c) Join up with the NDMC and start communicating.
(d) New school buildings and other educational facilities may be made safer by using hazard and risk maps in the planning stages.
(e) Direct all elementary school administrators to take precautions to keep school facilities safe and minimize any potential damage in the event of a natural catastrophe.
(f) In cyclone- and flood-prone locations, schools should be built with at least two stories.

(g) All elementary and mass education programs and institutions should include disaster preparedness into their curriculum.
(h) Conduct reaction and recovery exercises in collaboration with CPP every April and September, and raise awareness and educate instructors and students in high-risk locations on what to do during response and recovery operations.
(i) Repair and maintain structures in high-risk flood and cyclone zones.
(j) Build the Ministry's strategy for preventing and responding to potential disasters.
(k) For the Ministry's risk reduction efforts, design a sectoral emergency response system.
(l) Create a framework for the Ministry to share information about potential dangers.
(m) Make a backup strategy on how to handle disasters if it happens.
(n) Encouragement of school earthquake exercises, particularly in high-rise buildings (Dhaka, Chittagong, Rajshahi, Rangpur, Sylhet, and Mymensingh).

Emergency Response

Normal Times

(a) Assign a single Ministry Liaison Officer the role of "Focal Point" in Disaster Management.
(b) All elementary and secondary school curricula are updated in collaboration with DMB to incorporate disaster-related topics.
(c) Prepare teachers and students in affected regions for catastrophes by teaching them what to do in various emergency situations, and then mobilize them as volunteers to help with rescue, evacuation, and relief efforts in collaboration with local authorities.
(d) Be cautious to take the right BNBC precautions while building schools or doing repairs to primary school structures in cyclone- or flood-prone locations.
(e) Education institutions participating in the CPP should conduct large-scale drills for disaster preparation in April and September of each year to raise awareness and knowledge among the local population.
(f) When disaster strikes, local authorities should have access to primary schools and other facilities they control for use as shelters and relief centers.
(g) Coordinate with local authorities to recruit educators and students as rescue, evacuation, and relief volunteers.
(h) Make sure that alerts are being sent out in a timely and efficient manner across all relevant sectors.
(i) Maintain an efficient method of reporting, information sharing, and sector-wide communication.

Disaster Stage

(a) Put the requisite number of schools, together with their buildings, under the jurisdiction of the local government so that they may be used as a refuge and relief center in the case of a hurricane or flood.

(b) Set up a variety of schools to continue providing instruction during and immediately after natural disasters like floods and cyclones.

Rehabilitation Stage

(a) Assess the damage to educational facilities and provide restoration suggestions.

Directorate of Primary Education

The following tasks are in addition to the routine operations of this Division and its own work plan (contingency plan).

Risk Reduction

(a) In the event of a disaster, one Liaison Officer should be named as the focal point for management.
(b) All elementary schools and community-wide education programs should, in coordination with the Disaster Management Bureau, include disaster-related issues into their curriculum.
(c) In the event of a natural catastrophe, it is imperative that all elementary schools be built to serve as evacuation centers.
(d) Make sure your agency has a backup plan in place and that you always follow it.

Emergency Response

Normal Times

(a) Educators and students in affected regions should be given disaster response training and, if required, organized as volunteers in collaboration with local authorities and inspired to participate in rescue, evacuation, and relief efforts.
(b) Make sure that elementary school buildings in cyclone-prone locations are properly repaired and maintained.
(c) Education institutions participating in the CPP should conduct large-scale drills for disaster preparation every year in April and September to raise awareness and knowledge among the local population.
(d) Partner with the Disaster Management Bureau to integrate disaster education into K-12 and community-wide educational initiatives.

Disaster Stage

(a) Prepare for the possibility of severe cyclones and floods by putting the appropriate number of elementary schools under the jurisdiction of local authorities, so that they can serve as shelters and relief hubs if needed.

Rehabilitation Stage

(b) Find out how much damage elementary schools have sustained, and provide solutions for fixing them.

Ministry of Chittagong Hill Tract Affairs

Risk Reduction

(c) Appoint a high-ranking official to serve as the Ministry's main point of contact for disaster management.
(d) Advise all of its sections to provide their assistance to the relevant Ministries and Departments.
(e) Keep the PM Office, IMDMCC, and MoFDM in constant communication.
(f) Conduct a Ministry-wide analysis of potential threats and draw up a strategy for mitigating such threats.
(g) To reduce risks, the Ministry's action plan must be put into effect.
(h) Guarantee the Ministry allocates funds toward risk mitigation initiatives.

Emergency Response

Normal Times

(a) Carry out the plan to conduct a capacity development programme for the personnel on risk reduction and risk management, with a particular focus on earthquakes.
(b) Help the Ministry of Environment and Forest with their awareness campaigns on preventing environmental damage.
(c) The risk management programs have been allocated a certain amount of money.
(d) Get in touch with the Defense Ministry and see what arrangements can be made for evacuation and rescue.
(e) Prepare for a variety of emergency situations by establishing early warning and alarm systems, communication channels, a protocol for dealing with civil administration, and a list of available resources to use in relief and recovery efforts.
(f) Construct a backup strategy for the Ministry's risk mitigation and management programs.

Disaster Stage

(a) Help the local government out by doing damage and need assessments, providing medical care, and even airlifting injured people to safety. Also, provide relief supplies, medications, and medical personnel to the people who need them.
(b) Make sure that warnings are getting out there.
(c) Control spending to make sure it stays within the set budget.

Chittagong Hill Tract Development Board

In addition to normal functions, the CHDF will carry out the following functions.

(a) Maintain ecological harmony through safeguarding water supplies, mountain ranges, and other natural landmarks.
(b) Be careful to avoid destroying the slope and the mountains.

(c) Find out how susceptible CHT locations are to earthquakes and make a strategy to deal with them.

Ministry of Science and Information & Communication Technology

Risk Reduction

(a) Appoint a high-ranking official to serve as the Ministry's main point of contact for disaster management.
(b) Maintain attendance at the IMDMCC conference.
(c) Request that an evaluation of the current warning infrastructure be planned.
(d) Take an effort to locate a modern early warning system.
(e) It is imperative that the current warning technology be replaced with more up-to-date alternatives.
(f) Make a strategy and provide DMIC with the help it needs to create a cutting-edge early warning system.
(g) Create industry- and enterprise-wide plans for handling business interruptions.
(h) Integrate the Ministry's development planning procedures with disaster management concepts and best practices.
(i) Incorporate concerns about catastrophe risk management into the planning stages of any new industrial or manufacturing facilities.
(j) Create study plans for the Ministry of Hazardous Substances and Health and Environment.
(k) Always keep the NDMC updated on the status of sectoral risk reduction actions and programs.
(l) Create a method for reducing risks and being ready across industries.
(m) Put in place a Ministry of Risk Management and Communication risk communication system.
(n) Create a Ministry-wide emergency response mechanism.
(o) Get the Ministry to set up a framework for communicating risks across different industries.
(p) Make that the Ministry's risk management and mitigation efforts are adequately funded.
(q) Get together a backup plan for the organization.
(r) Examine Bangladesh's potential for earthquakes.
(s) Do not stop using the agency's backup plan.
(t) Distribute to GSB and BMD the agency readings obtained from Galmitre and other seismic monitoring equipment.

Emergency Response

(a) Ensure that warnings are being communicated effectively.
(b) Locate the alert messages that were sent out and direct the relevant departments toward the upgrades that will improve the technology's efficacy.
(c) Analyze the effects of risks at the regional and national levels of industry using a rigorous hazard and risk assessment.

Bangladesh Atomic Energy Commission

In addition to normal activities, the commission will perform the following duties.

(a) Keep an eye on any nuclear and radioactive supplies and machinery at hospitals, factories, labs, and other establishments to make sure they're secure.
(b) Take the necessary steps to deal with any nuclear or radioactive events that may occur.

Ministry of Youth and Sports

Risk Reduction

(a) Appoint a high-ranking official to serve as the Ministry's focal point for disaster management.
(b) The IMDMCC meetings need to have a high attendance rate.
(c) Find those willing to help in times of need, and provide them appropriate preparation and education. Add disaster management topics to the curriculum and provide disaster management courses for YTCs.
(d) Connect the DM&RD, MoFDM, and DMIC for a unified front.
(e) Conduct a Ministry-wide risk analysis.
(f) Make that a sectoral strategy to reduce risk is being developed.
(g) Provide for and properly allot funds in the budget.
(h) Facilitate Sector-wide learning and information sharing initiatives.
(i) Invest in your team's growth and development by giving them the tools they need to do their jobs well. Spend money on things like training, setting up lines of communication, and risk assessments.
(j) Make a plan to reduce risks and increase preparation across all sectors.
(k) Create a strategy for reducing and preparing for risks in your industry. Create a Ministry-wide emergency response mechanism.
(l) Put in place a Ministry of Risk Management and Communication risk communication system.
(m) The Ministry of Defense has to formulate a backup plan for its various risk mitigation and management initiatives.

Emergency Response

(a) Maintain contact with the emergency management data center (DMIC).
(b) Do what needs to be done to mobilize the Ministry's resources, such as its trained volunteers, to aid in replies.
(c) Get the word out to the right people, especially DMIC, by preparing and distributing situation reports.
(d) Take up certain roles as outlined in the industry's backup strategy.
(e) If required, it should reorganize its resources, including its personnel.

Department of Youth Development

Besides the normal function the department will carry out the task on disaster management.

(a) Make sure that the department is well-represented on various disaster management boards.
(b) Get youth groups involved voluntarily in catastrophe prevention and relief efforts.
(c) Promptly provide instructions to regional offices to guarantee youth involvement in disaster management.
(d) Ensure that all youth training centers provide disaster management training, either by incorporating it into existing curricula or by creating entirely new programs.

Ministry of Religious Affairs

Risk Reduction

(a) Appoint a high-ranking official to serve as the Ministry's main point of contact for disaster management.
(b) The IMDMCC meetings need to have a high attendance rate.
(c) Get in touch with the MoFDM and the DMIC.
(d) Make sure a strategy for lowering risks in certain industries is developed.
(e) Don't cut down on funding for safety measures.
(f) Create a strategy for handling potential threats and preparing for emergencies in each individual industry.
(g) Make sure religious leaders are involved in developing a strategy to lessen the likelihood of devastating natural disasters, and provide them the tools they need via training on disaster preparedness and community resilience so that they can spread the word through local churches and synagogues.
(h) Provide for and properly allot funds in the budget.

Emergency Response

Normal Times

(a) Plan sector-wide events to raise awareness and disseminate information.
(b) To increase public awareness, particularly in disaster-prone regions, it is important to form strategic partnerships with local religious leaders and non-governmental organizations.
(c) Make sure there's enough money in the budget to carry out things like training, danger and threat assessments, effective communication, and increasing employee capability and knowledge.
(d) The Islamic Foundation may compile a list of interested Imams and other religious leaders, and then give them with training on disaster preparation, focusing on the non-structural vulnerabilities of earthquakes.

(e) The Ministry of Defense has to formulate a backup plan for its various risk mitigation and management initiatives.

Warning/Alerting Stage

Mobilize the trained volunteers to:

(a) Maintain contact with the emergency management data center (DMIC).
(b) Do what needs to be done to mobilize the Ministry's resources to aid in responding.
(c) Get the word out to the right people, especially DMIC, by preparing and distributing situation reports.
(d) Take on designated tasks in accordance with SOD and the sectoral strategy.
(e) If required, it should reorganize its resources, including its personnel.

Ministry of Cultural Affairs

Risk Reduction

(a) Appoint a high-ranking official to serve as the Ministry's main point of contact for disaster management.
(b) The IMDMCC meetings need to have a high attendance rate.
(c) Get in touch with MOFDM and DMIC, and start talking.
(d) Engage NGOs, CBOs, and UP in a sector-wide risk assessment and design an action plan for mitigating risks and being ready for disasters with an eye on using cultural events as a means of raising awareness and educating the public about DRR and CC.
(e) Prepare a risk management and emergency response strategy with the help of civil society groups.
(f) Implement disaster management strategies into the Ministry's planning initiatives.
(g) Make sure money is set out in the budget for risk mitigation efforts.

Emergency Response

Normal Times

(a) Create sector-wide opportunities for education and awareness-raising about disaster preparedness and risk.
(b) Dedicate funds to a variety of initiatives aimed at enhancing the skillsets of employees, raising public awareness, facilitating efficient communication, and analyzing potential risks.
(c) The Ministry of Defense has to formulate a backup plan for its various risk mitigation and management initiatives.
(d) Make sure that non-governmental organizations, community-based organizations, Pourashava, and the United People are involved in disaster-related public awareness campaigns.

(e) Create a partnership with the print and digital media in the process of raising awareness. Create sector-wide opportunities for education and awareness-raising about disaster preparedness and risk.
(f) Dedicate funds to a variety of initiatives aimed at enhancing the skillsets of employees, raising public awareness, facilitating efficient communication, and analyzing potential risks.
(g) The Ministry of Defense has to formulate a backup plan for its various risk mitigation and management initiatives.
(h) Make sure that non-governmental organizations, community-based organizations, Pourashava, and the United People are involved in disaster-related public awareness campaigns.
(i) Create a partnership with the print and digital media in the process of raising awareness.

Alerting Stage

(a) Maintain contact with the emergency management data center (DMIC).
(b) Involve and organize the community warning networks of NGOs, CBOs, Pourashava, and UP.

Disaster Stage

(a) Do what needs to be done to mobilize the Ministry's resources to aid in responding.
(b) The company should reorganize its resources, including its employees, if this becomes required.

Ministry of Civil Aviation and Tourism

This Ministry, in addition to its regular responsibilities, will also carry out the following.

Risk Reduction

(a) Establish a Ministry official as the department's primary contact for disaster management.
(b) Instruct all departments and bureaus to collaborate on a strategy to reduce the likelihood of catastrophe occurring.
(c) Maintain constant communication with the Department of Agriculture and Emergency Management and see that your own strategy is developed.
(d) Conduct a comprehensive risk analysis of all of the country's most vital infrastructure, storage facilities, airliners, and tourism hotspots.
(e) Taking precautions will make current infrastructure less susceptible to damage from natural disasters and will make services and systems more robust overall.
(f) Develop a strategy for mitigating risks in your industry and put it into effect.
(g) Make sure there's money in the budget for its risk mitigation efforts.
(h) Protect the travelers and sightseers by using risk avoidance strategies.
(i) Instruct all departments and offices to work together and devise a comprehensive strategy for mitigating potential dangers.

(j) Educate and raise disaster preparedness awareness among your personnel.
(k) Make that a post-impact hazard and risk analysis is conducted and that risk treatment measures are included into the restoration process.

Emergency Response

Normal Times

(a) During the effect of a danger, services should remain operational, and thus preparation is essential.
(b) Under the direction of the National Disaster Management Council and the Inter-Ministerial Disaster Management Coordination Committee, arrange for air transport to evacuate critically wounded people from active airports.
(c) Provide all necessary assistance to the local government throughout rescue, evacuation, relief, and restoration efforts.
(d) The Ministry of Food and Disaster Management may, if required, make use of good offices to acquire aircraft from the Flying Club, etc., for use in disaster relief operations.
(e) Facilitate the prompt arrival of international aid planes by clearing the airspace quickly.
(f) Airports should provide enough parking and takeoff and landing areas for aircraft participating in rescue and relief efforts.
(g) Create a plan on how to charge for the landing and transportation of aircraft on aid missions from countries that are friendly to yours.
(h) Move the at-risk air traffic services somewhere secure.
(i) Rapidly fix the damaged airports.
(j) Oversee the maintenance of adequate fuel and replacement parts supplies at all airports, civil aviation hubs, and tourism infrastructure facilities, with particular attention paid to locations prone to natural disasters.
(k) Obtain planes for disaster assistance missions following MoFDM's request.
(l) Landing and flight costs should be levied on foreign aircraft participating in humanitarian activities, and the process for collecting them should be outlined.

Disaster Stage

(a) As instructed by the IMDMCC/NDMC, provide air evacuation services.
(b) Assist the International Catastrophe Management Coordination Center/National Disaster Management Center (IMDMCC/NDMC) in air evacuations of disaster victims, with a focus on those who have suffered significant injuries.
(c) Quickly authorize international aid flights to land so they may unload their supplies.
(d) Make sure there's adequate room for rescue planes to land.
(e) Safeguard infrastructure and maintain service continuity by managing them under the effect of a danger.

Rehabilitation Stage

(a) Prepare gasoline and other petroleum items that may be needed for emergency response, relief, and recovery efforts.
(b) Clear the way for international aid planes to fly in safely.
(c) Provide landing and takeoff facilities for aircraft involved in rescue and relief efforts.
(d) Take action to restore damaged airports so they can serve passengers after a disaster.
(e) Plan for the safety of infrastructure and the continuation of services during the effect of the danger.
(f) Conduct a damage and loss analysis.
(g) If infrastructure maintenance on a longer time scale is required, it should be done.

Civil Aviation Authority of Bangladesh

In addition to normal activities, the authority will perform the following duties.

(a) Select one Liaison Officer to serve as the authority's main point of contact for all disaster-related matters under Disaster Management.
(b) Protect the integrity of the air navigation system at all times, and have a contingency plan for personnel and supplies ready in case of an emergency.
(c) The airports should all have their own backup plans ready.
(d) Make sure the airport and runway (cargo handling, warehouses) are prepared to assist the disaster response program as quickly as possible following a catastrophe (search-and-rescue personnel, equipment, etc.).
(e) During times of crisis, it was easier to transport necessary supplies and gear for search-and-rescue operations.
(f) After receiving early warning and signals, notify all involved parties and aircraft.

Ministry of Land

In addition to normal activities, the Ministry will perform the following duties.

Risk Reduction

(a) Appoint a high-ranking official to serve as the Ministry's main point of contact for disaster management.
(b) Participate at every IMDMCC gathering.
(c) Hazard maps may be used to create a strategy for mitigating risks and preparing for emergencies in different industries.
(d) Get the Land Ministry involved in the risk assessment and mitigation process.
(e) Plan for and zone for potential disasters, and make disaster risk reduction a priority in your community's overall development strategy.
(f) Create a plan for settling char and khas territory.
(g) Get the Ministry to set up a framework for communicating risks across different industries.

(h) The Ministry of Defense has to formulate a backup plan for its various risk mitigation and management initiatives.
(i) Integrate disaster management within the Ministry's policymaking and planning processes.
(j) Don't cut down on funding for safety measures.
(k) Set aside funds to fund initiatives like training and education programs for employees, setting up solid channels of communication, and analyzing potential dangers.

Emergency Response

Normal Times

(a) Plan sector-wide events to raise awareness and disseminate information.
(b) Raise public awareness of river erosion as a result of dangers by forming strategic partnerships with electronic media, NGOs, and CBOs.
(c) Get in touch with the MoFDM and the DMIC.

Disaster Stage

(a) Make sure you get in touch with DMIC and keep in touch with them.
(b) Do what needs to be done to mobilize the Ministry's resources to aid in responding.
(c) Get the word out to the right people, especially DMIC, by preparing and distributing situation reports.
(d) Take on designated tasks in accordance with SOD and the sectoral strategy.
(e) If required, it should reorganize its resources, including its personnel.
(f) When help from the Ministry of Water Resources is needed, take the appropriate action.
(g) Maintain open lines of communication, accurate reporting, and timely information during times of crisis.

Ministry of Foreign Affairs

The Ministry, in addition to normal responsibilities, will perform the following duties.

Risk Reduction

(a) Appoint a responsible senior member of staff to serve as the Focal Point and ensure regular communication with the NDMAC, MoFDM, IMDMCC, and NDMC.
(b) Help promote the government's efforts to reduce catastrophe risks and its successes in doing so in international forums.
(c) It is important for Bangladesh to keep in touch with international governments and donors (such as NGOs) via its overseas embassies.
(d) Help the NDMC, MoFDM, and IMDMCC through the process of receiving relief and recovery aid from a foreign government or agency by giving them technical guidance and assistance.

(e) Help other departments create strategies for receiving aid from international organizations after a natural catastrophe.
(f) Set up protocols and mechanisms within Bangladesh's diplomatic representations abroad to guarantee that aid for relief and recovery is delivered on time by governments and non-governmental organizations from other countries.

Emergency Response

Normal Times

(a) One Liaison Officer will be designated by the Ministry as the primary contact for disaster management.
(b) Direct all Bangladeshi diplomatic missions to actively pursue connections with international Red Cross and Red Crescent organizations.
(c) Make that the Bangladesh Missions in Geneva, Brussels, and New York are aware of the catastrophe situation in the nation and are prepared to brief prospective sources of help and support.
(d) Encourage Bangladesh's diplomatic representations overseas to maintain constant and open lines of communication with DHA and the International Federation for Red Cross and Red Crescent Societies (IFRCS).

Alert and Warning Stage

(a) Keeping the Bangladesh Missions in Geneva, New York, and Brussels apprised of the catastrophe situation in the nation and ready to brief any potential aid/assistance source is essential.

Disaster Stage

(b) Follow the GoB's/instructions NDMC's for the submission of requests for international assistance to foreign governments and agencies.
(c) Cooperate closely with the NDMC and the MOFDM/IMDMCC and advise them on the process for making a request to a funding source.

Rehabilitation Stage

(a) Maintain regular communication with the Ministry of Home Affairs and the Ministry of Disaster Management to streamline the visa application process for international aid workers.
(b) It is important to keep foreign governments, international donor organizations, and other contributors updated on how their money is being used.
(c) Make sure a letter of thanks is sent to each donor nation, either via their embassy in Dhaka or through our embassy in their country.
(d) Actions should be taken to ensure that fishermen and others who are displaced across international borders as a result of cyclones, storm surges, or any other catastrophe are able to safely return home.

Ministry of Textiles and Jute

The Ministry, in addition to normal responsibilities, will perform the following duties.

Risk Reduction

(a) Appoint a high-ranking official to serve as the Ministry's main point of contact for disaster management.
(b) The IMDMCC meetings need to have a high attendance rate.
(c) Get in touch with the MoFDM and the DMIC.
(d) Conduct a sector-wide risk analysis, and draft a strategy for mitigating threats and enhancing readiness.
(e) A risk reduction strategy needs input from both Ministry departments, so make sure they're included in the process.
(f) Implement disaster management strategies into the Ministry's planning initiatives.
(g) Make sure money is set out in the budget for risk mitigation efforts.
(h) Use the Ministry's strategy for preventing and dealing with potential disasters.

Emergency Response

Normal Times

(a) Plan sector-wide events to raise awareness and disseminate information.
(b) Join forces with non-governmental organizations to educate those who work in the textile and jute industries.
(c) Set aside funds to fund initiatives like training and education programs for employees, setting up solid channels of communication, and analyzing potential dangers.
(d) The Ministry of Defense has to formulate a backup plan for its various risk mitigation and management initiatives.

Disaster Stage

(a) Make sure you get in touch with the DMIC and keep it going.
(b) Do what needs to be done to mobilize the Ministry's resources to aid in responding.
(c) Get the word out to the right people, especially DMIC, by preparing and distributing situation reports.
(d) Take on designated tasks in accordance with SOD and the sectoral strategy.
(e) If required, it should reorganize its resources, including its personnel.

Ministry of Commerce

In addition to normal functions, the Ministry will perform the following duties:

Risk Reduction

(a) The Ministry will gather data from regional offices to keep tabs on the availability and cost of necessities.
(b) Create a Ministry of Defense readiness strategy plan for mitigating risks in certain industries.

(c) Create a framework for sharing information about potential dangers inside and across government departments.
(d) Prepare a sectoral response strategy for emergencies.
(e) Ensure the Ministry's financial resources are adequate.

Emergency Response

Normal Times

(a) Appoint one Ministry Liaison Officer to serve as the Department's main point of contact for disaster management.
(b) Get a backup plan together for each individual industry.

Alert and Warning Stage

(a) It is important to provide advance notice to those who need to know and to take whatever necessary precautions to ensure the safety of one's own personnel and stored products in locations where a catastrophe is imminent.

Rehabilitation Stage

(a) Planning the import of CI sheet, cement, etc., for home repairs, and fertilizers and seeds, if needed, for agricultural product production, according to the needs of the relevant Ministry/Agency.
(b) Get ready for the emergency import of lifesaving supplies and medications.
(c) Keep the impacted regions supplied with essentials at reasonable prices.
(d) Make sure the individuals in the impacted regions may still import the supplies they need.
(e) Staff from the Ministry and its Directorates will monitor commodity prices, supplies, and distribution in disaster zones, reporting their findings regularly.
(f) In addition to grains, ensure the availability of other essential food items such as salt, legumes, vegetables, edible oil, milk, and potatoes.

Ministry of Power, Energy, and Mineral Resources

The Ministry (including subsidiary agencies) will carry out the following responsibilities in addition to enforcing its own Action Plans (contingency plans).

Risk Reduction

(a) Appoint a single competent official within the Ministry to act as the Department's primary contact for disaster management.
(b) Distribute preemptive detailed instructions on roles and responsibilities to relevant personnel in the event of a crisis and its aftermath.
(c) To reduce risks in certain industries, the Ministry needs a strategy, which you may help them create.
(d) Make sure the Ministry conducts a risk assessment of its disaster management efforts on a sectoral level.
(e) Protect the Ministry's ability to allocate resources toward risk mitigation and management in the annual budget.

(f) Conduct and oversee risk reduction initiatives to lessen the susceptibility of services and systems and increase the robustness of physical infrastructure.
(g) Preparation ahead of time will help keep services running smoothly in the face of a disaster.
(h) Educate and raise the profile of the personnel.
(i) Perform a power plant risk analysis.
(j) Maintain adequate supplies of replacement parts and other commodities at the Chittagong, Comilla, Barisal, Khulna, Iswardi, Bogra, and Rangpur distribution centers of the Power Development Board and the Rural Electrification Board.
(k) Build the Ministry's strategy for preventing and responding to potential disasters.
(l) Create a method for rapid reaction in certain sectors in the event of an emergency.
(m) Create a network of ministerial and interministerial channels for communicating on risk reduction.
(n) Make a backup strategy for the Ministry's efforts to mitigate potential disasters.
(o) Make public the agency's seismic emergency procedure.
(p) Raise everyone's consciousness of what to do in the event of an earthquake.
(q) The organization should do everything it takes to prepare its whole personnel for emergencies.
(r) In badly damaged locations, keep a database to estimate the amount of electrical equipment and hazardous substances available.
(s) Verify that all recently built structures have earthquake safety equipment installed and/or earthquake safety procedures implemented.
(t) Avoid potential disasters by installing automatic shutoff systems at all electric and gas transmission facilities.
(u) Be careful to conduct geological surveys and draw up risk maps to eventually cover the whole nation.

Emergency Response

Normal Times

(a) Some precautions, in addition to the regular tasks and obligations of the departments under management, must be maintained ready to ensure that the power plant, the supply/distribution of power, and the power system are not adversely affected. Important facilities (such as hospitals, radio/television stations, civic and social facilities) must be kept running during repairs to damaged power systems, and this can only be done if the power supply is restored as soon as possible.
(b) Provide staff with comprehensive guidance on their roles throughout the pre-, during-, and post-disaster phases of natural catastrophes.
(c) Schedule frequent disaster-related workshops and on-the-job training for all staff members working in high-risk locations.
(d) Maintain contact with the emergency management data center (DMIC).

Alert and Warning Stage

(a) Protect power plants in remote places from the potential dangers of tidal bore and flood floods by putting together a plan of action now. Arrange for all of the central stores of the Power Development Board/Rural Electrification Board in Chittagong, Comilla, Barisal, Khulna, Iswardi, Bogra, and Rangpur to stock spare parts for 132 kV towers, 33 kV tower/Poles, 11 kV poles, LT poles, line conductor of different size; transformer of different size; and running diesel generator sets.
(b) When a disaster warning signal is sent, such as for a hurricane or a flood, it is important to set up a central location from which local authorities may be reached without interruption.
(c) Appoint one person to serve as the Liaison Officer, who will liaise between the Dhaka headquarters, the Local Disaster Management Committee, and the many damaged stations.
(d) The lines of communication must be kept open at all times.
(e) Ensure that all employees and police are evacuated to the closest safe location.
(f) Prepare the people, transportation, tools, supplies, and backup power line by setting up appropriate illumination.
(g) If a conductor wire were to crack, it would be necessary to immediately turn off electricity at the closest power plant or substation to avert an accident.
(h) Form an ongoing central Disaster Management Committee and establish stations in various regions affected by the disaster.

Disaster Stage

(a) Take shifts in the Control Room (24 h).
(b) Make sure that the resources, such as repair tools, transformers, and so on, required to reinstall broken lines and make repairs are readily available.
(c) Immediately restore electricity to the impacted locations. Dhaka Headquarters should be notified of the need for further reinforcement.
(d) Do what needs to be done to mobilize the Ministry's resources to aid in responding.
(e) If required, it should reorganize its resources, including its personnel.
(f) It is imperative that immediate repairs be made to the infrastructure in order to maintain supplies.
(g) Control supplies of gasoline and other petroleum products needed for disaster response, relief, and recovery.

Rehabilitation Stage

(a) Keep working toward restoring electricity and getting life back to normal.
(b) Assess the harm or loss that has occurred.
(c) In the aftermath of a catastrophe, it's important to communicate the amount of money needed to restore electricity and rebuild to the appropriate offices and authorities so that they can provide the necessary funding.

(d) Project ideas should be planned and prepared for reinstallation or rehabilitation and submitted to the relevant authorities.
(e) Get the word out to the right people, especially DMIC, by preparing and distributing situation reports.
(f) Whenever required, invest in extensive infrastructure repair.

Petroleum and Mineral Resources Division

Along with its regular tasks, this department will also coordinate the distribution of fuel to regions hit by calamities like cyclones and tidal bores via organizations like the Bangladesh Petroleum Corporation.

Risk Reduction

(a) Select a reliable Division employee to serve as the Focal Point for Disaster Management.
(b) Bangladesh Petroleum Corporation Risk Assessment and Mitigation Strategy Development (BPC).
(c) Conduct and oversee risk reduction initiatives to lessen the susceptibility of services and systems and increase the robustness of physical infrastructure.
(d) Preparation ahead of time will help keep services running smoothly in the face of a disaster.
(e) Educate and raise the profile of the personnel.
(f) Create a method for rapid reaction in certain sectors in the event of an emergency.
(g) Create a network of ministerial and interministerial channels for communicating on risk reduction.
(h) The Bangladesh Petroleum Corporation (BPC) has to establish a backup plan for its risk management procedures.

Emergency Response

Normal Times

(a) If there is any problem with the petroleum distribution station or pipeline, the Bangladesh Petroleum Corporation will fix it.

Alert and Warning Stage

(a) In the event of a natural catastrophe, the Bangladesh Petroleum Corporation (BPC) will notify its oil marketing businesses, agency/dealers, and anybody else involved in the distribution of petroleum products.
(b) In particular, make sure that distribution centers and stations in and near potentially impacted regions have enough supplies of kerosene, gasoline, and other petroleum products.
(c) If there is less stock than normal at the aforementioned distribution centers/stations, or if there is a risk that stock may be depleted quickly, the Bangladesh Petroleum Corporation will move quickly to establish arrangements for reimbursement.

Disaster Stage

(a) In the event of an emergency, the Bangladesh Petroleum Corporation will ensure that all distribution centers and stations in and surrounding the affected regions have access to petroleum supplies.
(b) Coordinate the availability of POL and its supply and distribution to affected persons by staying in constant contact with fuel oil firms and suppliers of petroleum products in impacted regions.

Rehabilitation Stage

(a) After the crisis has ended, the aforementioned actions will still be taken.
(b) Officers from Bangladesh Petroleum Corporation, its marketing partners, and anyone else with a stake in the aftermath of a catastrophe should take an active role in coordinating and carrying out the distribution of fuel and other supplies.
(c) Until the end of the relief activities, they will be in constant contact with the government. Oil industry leaders and authorities will be active participants in both regional and local DMCs.

Geological Survey of Bangladesh (GSB)

The following are GSB's responsibilities in regard to earthquake preparedness in Bangladesh.

Risk Reduction

(a) Make sure you choose a reliable member of the GSB staff to be the main contact in times of crisis.
(b) Assess potential dangers and draw out a strategy to lessen them at the Geological Survey of Bangladesh (GSB).
(c) Conduct and oversee risk reduction initiatives to lessen the susceptibility of services and systems and increase the robustness of physical infrastructure.
(d) Prepare for the effects of a disaster and keep services running smoothly during and after it occurs.
(e) Promote disaster preparedness education and awareness among GSB personnel.
(f) Create a framework of coordinated assistance from different industries in the event of a disaster.
(g) Create a network of ministerial and interministerial channels for communicating on risk reduction.
(h) Create a backup plan for the GSB's risk mitigation initiatives and share it with the agency.
(i) Construct risk maps, covering in sequence all major cities and ports at danger from earthquakes.
(j) Distribute earthquake risk maps to the appropriate agencies and make the required preparations for doing so.
(k) Create a Geological Centre to coordinate earthquake risk assessments with national and international organizations such as the Bangladesh Meteorological Department (BMD), the Bangladesh Atomic Energy Research Commission

(BAERC), the Bangladesh University of Engineering and Technology (BUET), and others.

(l) Your help is needed for the Ministry of Housing and Public Works and its sub-departments, the municipal development authority, and other connected organizations and institutions to establish sound policy on earthquake-resistant building construction.
(m) Help enhance the automated shutoff mechanism used by electric and gas transmission agencies and enterprises.
(n) Determine the earthquake-prone structures in your area and assess their geological position.
(o) To learn how earthquakes will affect different areas of the nation, it is necessary to study the geology and structure of the land across the board.
(p) Get the geological mapping program done up to at least a 1:50 k scale for the whole nation, and a 1:5,000 scale for urban areas up to the District level. The map will reveal the location and features of several geological units.
(q) Better depictions of the country's active and dormant faults and tectonic lineaments are needed on updated structural and tectonic maps.
(r) Prepare a new seismic and tsunami zoning map for Bangladesh utilizing the most recent data, technology, and understanding.
(s) Make a difference in the formulation of policies about the risk of earthquakes.
(t) Help raise public understanding by holding seminars, workshops, and other educational events, and cooperating with other organizations doing the same.
(u) Immediately after an earthquake, take into account geological, engineering, socioeconomic, and other types of loss in order to prepare a map of the earthquake's strength and damage.

Emergency Response

Normal Times

(a) The GSB will watch over its facilities to make sure nothing bad happens to them.

Alert and Warning Stage

(a) In the event of a catastrophe, GSB will notify its affiliated agencies and any other relevant organizations.

Department of Explosives

In addition to normal activities, the department will perform the following duties.

(b) To be ready for industrial catastrophes, it is important to keep an eye on the transfer of dangerous products including explosives, corrosives, inflammable liquid gas, and acids.
(c) Make sure that secondary risks and catastrophes are dealt with appropriately in the event of liquid spills, gas emissions, or explosions.
(d) Before breaking or dismantling condemned ship facilities and industries, a gas-free certificate must be issued.

Ministry of Labor and Employment

Risk Reduction

(a) Appoint a high-ranking official to serve as the Ministry's main point of contact for disaster management.
(b) The IMDMCC meetings need to have a high attendance rate.
(c) Get in touch with MOFDM and DMIC, and start talking.
(d) Conduct a risk analysis of the industrial labor sector and develop a plan to mitigate any issues found.
(e) Create a standard to enhance workplace safety and reduce risks for employees.
(f) Consider implementing a plan to pay and insure employees who must operate in dangerous environments.
(g) Make sure there's enough money in the budget to do things like train and educate employees, set up efficient lines of communication, and do risk assessments.

Emergency Response

Normal Times

(a) Build a strong alliance with the relevant government agencies.
(b) Organize events to raise workers' consciousness and teach them about safety precautions in the workplace.
(c) The Ministry of Defense has to formulate a backup plan for its various risk mitigation and management initiatives.

Disaster Stage

(a) Make sure you get in touch with the DMIC and keep it going.
(b) Do what needs to be done to mobilize the Ministry's resources to aid in responding.
(c) Get the word out to the right people, especially DMIC, by preparing and distributing situation reports.
(d) Take on designated tasks in accordance with SOD and the sectoral strategy.
(e) If required, it should reorganize its resources, including its personnel.

Ministry of Liberation War Affairs

Risk Reduction

(a) Appoint a competent Ministry official to serve as the department's main point of contact in times of disaster.
(b) Maintain attendance at IMDMCC and other applicable committee meetings.
(c) Facilitate open lines of communication between MoFDM and DMIC.
(d) Conduct a risk analysis and develop a strategy to mitigate threats and improve emergency readiness for the Ministry.
(e) Don't cut down on funding for risk management measures.

Emergency Response

Normal Times

(a) Do things to raise people's consciousness and teach them new things in many fields.
(b) Create a long-term alliance with relevant government agencies.
(c) Implementing several initiatives aimed at enhancing the team's abilities, including training, improved communication, and a thorough assessment of potential dangers.
(d) Create a backup strategy for the Ministry's risk management and mitigation efforts.

Disaster Stage

(a) Get in touch with the DMIC and keep it going!
(b) Do what needs to be done to mobilize the Ministry's resources to aid in responding.
(c) Create incident reports and make sure they go out to the right people, like DMIC.
(d) Carry out designated tasks in accordance with SOD and the sectoral strategy.
(e) If required, it should reorganize its resources, including its employees.

Ministry of Expatriates' Welfare and Overseas Employment

Risk Reduction

(a) Appoint a competent Ministry official to serve as the department's main point of contact in times of disaster.
(b) Please make sure that people are attending the IMDMCC sessions.
(c) Connect MOFDM and DMIC to provide efficient communication.
(d) Conduct risk assessments of the international labor market and the potential for natural disasters and produce a strategy for mitigating those risks and preparing for them to be implemented by the Ministry and relevant agencies.
(e) Don't cut down on funding for risk management measures.

Emergency Response

Normal Times

(a) Do things to raise people's consciousness and teach them new things in many fields.
(b) To help employees prepare for the challenges they may face in their host countries and get perspective on the opportunities they will encounter, it is important to establish strategic partnerships with Overseas Employment Agency.
(c) Conducting a variety of actions aimed at increasing the team's capabilities, including training, communication, and the identification and assessment of potential risks.
(d) Keep in touch with the Ministry of Foreign Affairs to provide particular information on disaster-stricken countries and Bangladeshi workers.

(e) Create a backup strategy for the Ministry's risk management and mitigation efforts.

Disaster Stage

(a) Create a database with information on your foreign workers that can be shared with loved ones and relief organizations in the event of an emergency.
(b) Maintain open lines of contact with the foreign ministry and make preparations for their aid in the event of a tragedy in any of your abroad postings.
(c) Make sure you get in touch with the DMIC and keep in touch with them.
(d) Do what needs to be done to mobilize the Ministry's resources to aid in responding.
(e) Create incident reports and make sure they go out to the right people, like DMIC.
(f) Carry out designated tasks in accordance with SOD and the sectoral strategy.
(g) If required, it should reorganize its resources, including its employees.
(h) Responsible for managing and allocating Ministry money and resources in accordance with established budgetary guidelines.

Other Ministries

If a national emergency is declared, all government agencies are obligated to do everything they can to prevent more destruction, minimize losses, and get life back to normal as soon as possible. Where the SOD does not specify roles and responsibilities, those Ministries will develop and implement their own action plan on risk reduction and response management in accordance with section No 4.1 General Roles and Responsibilities of all Ministries, Divisions, Departments, and Government Owned Corporations.

Within governments and between governments and other essential parties including funders, NGOs, and the corporate sector, coordination has undergone tremendous change during the last decade. Separating food and disaster management at the Ministry level, creating the Ministry of Disaster Management and Relief as the coordinating national agency for disaster risk reduction and emergency management at the policy level, and establishing the Department for Disaster Management to coordinate national disaster management initiatives are all steps forward in the effort to improve the system.

In its response to catastrophes from 2001 to 2012, the Ministry of Disaster Management and Relief encountered difficulties and gained some insights as a result of its experiences. For instance, the Department of Disaster Management was supposed to efficiently run humanitarian assistance, recovery, and reconstruction programs, but this has not happened because the Disaster Management Act 2012 has only passed through parliament and not all of the necessary procedures and arrangements have been finalized. The government of Bangladesh has already established a number of regulations and standards to aid with disaster response programs, but there were still many more that hadn't been approved as of 2012. As a result of this void, local authorities were sometimes forced to make ad hoc judgments with wide variations. Both government and non-government organizations must adhere to consistent standards, and in cases when they may need to use discretion, local authorities must be provided with explicit guidance.

Humanitarian organizations struggled to coordinate their efforts after a catastrophe because they each developed their own response plan with goals that were too dissimilar to work together. Fewer storm shelters were available than needed. On top of that, the Department of Catastrophe Management has trouble getting enough trained staff mobilized in the event of a disaster. It was also found that the mechanism for storing, compiling, and analyzing data was insufficient, and that local disaster management committees lacked awareness and expertise in need assessment. The District and Upazila Disaster Management Committees lacked political leadership.

There were central government representatives entrusted with communicating with local disaster management committees. These representatives were civil servants from Bangladesh. For instance, the Deputy Commissioner heads the District Disaster Management Committee, but the Upazila Nirbahi Officer is in charge of the Upazila Disaster Management Committee. Managers and leaders of Disaster Management Committees may lack expertise in disaster management, but they have been given the responsibility of coordinating and managing disaster risk reduction initiatives in accordance with the Disaster Standing Orders. Because local political leadership is not engaged in catastrophe risk reduction, there is no reflection of people's interests or feeling of responsibility. A lack of transparency on the part of Bangladesh's Upazila and District governments leaves disaster-prone citizens in the dark regarding the nature of the disaster risk reduction initiatives being developed and put into action in their communities. However, the Union Disaster Management Committee's chairman has only existed on paper, and the committee's other members lack adequate training in disaster management protocol.

As the Upazila Nirbahi Officer and the Upazila Project Implementation Officer are in charge of the government's assigned resources for disaster risk reduction, the Union level Disaster Management Committee faces significant obstacles to its effectiveness. While under the watchful eye of Union Parishad, they are tasked with coordinating and ensuring the allocation of resources for DRR implementation. In practice, the Chairman, who is supposed to represent the people who live in the area governed by the Union Parishad, is not consulted on how funds should be distributed. Currently, it is crucial to train members of Disaster Management Committees on need assessment, create volunteer teams in neighborhoods to swiftly mobilize human resources for evacuation, rescue, and relief distribution, and generally improve communities' disaster preparedness through community-based disaster risk reduction programs. In addition, political leadership is essential to advance DRR for mitigating the root causes of catastrophes in Bangladesh.

A well-established coordination mechanism for the Government and development partners and other players was possible because of the 2012 reform of the disaster management system. It should be noted that not all of the coordination committees indicated in the SODs are currently active [11]. They have been dissolved or never existed to begin with and have been replaced by other committees. No evaluation of their effectiveness has been made public. The coordinating structure for disaster management places a premium on rapid response. It is harder to coordinate efforts to lessen danger, beef up preparation, and speed up recuperation. This is because the government's operations are spread over many distinct Ministries, and the development partners' activities are spread across several different ministries.

When it comes to natural disasters, Bangladesh is among the most vulnerable nations. Repeated calamities devastate the resource base of the impoverished and diminish their economic potentials. Proper disaster management planning and integrating disaster management efforts with local and national development plans might reduce catastrophe effects and vulnerabilities. The purpose of this research is to determine how Union Disaster Management Committees (UDMCs) may be improved so that they are more effective in carrying out Standing Orders on Disaster (SOD) [12].

The UDMC is not well-versed in the SOD. Most of them operate in the conventional fashion rather than the pecking order. Some UDMC participants were found to be less engaged than others in efforts to educate the public and train first responders after the Hood disasters. The Union Disaster Management Committee (UDMC) may be an integral part of community-level preparations for disasters and disaster relief. Standing Orders on Disaster (SOD) released by the Government of Bangladesh outline the functions and duties of UDMCs in detail. A lack of awareness/knowledge, a lack of resources, and the need to enhance UDMCs are all true reasons why most UDMCs remain dormant. Members of the UDMC should get training on disaster preparedness in their communities. Standing Orders on Disaster (SOD) at the Union level need quick attention to political will, resource allocation, the construction of a robust monitoring and follow-up mechanism, and the Disaster Management Act.

Simply having two women on each DMC is insufficient to fully reflect women's demands and abilities. Unfortunately, the question of whether or not women may actively engage in and shape the DMCs remains untested and unanalyzed [13]. There is no road map for how to best assist and strengthen DMCs. It is not able to obtain this data or to identify where committees are working and where they are not, despite the fact that there are a lot of actions to enhance the capacity of the Disaster Management Committees. All DRR initiatives at the field level where resources are given at the national level via district administration include elected local government members at the Union and Upazila levels. However, the amount of given resources is insufficient to back the activities in this very populated nation with many threats impacting the people round the year.

The field of disaster management is becoming more dominated by human-caused and technical catastrophes. It cannot be denied that the private sector plays an important role and is actively involved in the Disaster Management Committee at a variety of levels. This issue in our nation goes beyond the "charity" mentality of the business community. Harmful consumption, production, distribution, and antisocial behaviors, mindsets, and practices must be minimized. However, the catastrophe and climate change condition may be improved by the development and dissemination of novel goods and services. The third problem is choosing between near-term and far-term gains, advantages, strategies, and plans. While residents prefer to focus on the here-and-now, professionals and experts consistently advocate for the future. As a result, the Disaster Management Committee needs to schedule frequent meetings to address the finance problem facing DRR programs.

When it comes to disaster preparedness, Bangladesh is light years ahead of most other nations. This is why the nation is held up as an example of how to respond to natural calamities. Our assets and traits include things like a strong community, active volunteers, an early warning system, community-based decision-making, government commitment, a thriving non-governmental organization sector, and a commendable legal and institutional structure. Our research shows that greater effort and dedication are needed to assess the legal and policy directives' provisions' actual application. DRR operations can only go forward with the help of strong Disaster Management Committees.

This chapter's goal was to educate the reader about the various national and local disaster management committees required by the SOD so that they may acquire comprehensive information about their responsibilities in disaster risk reduction. This chapter detailed the establishment of the committee and its subsequent deliberations. The report also made an effort to demonstrate how the combined efforts of many committees may strengthen national initiatives to reduce catastrophe risk.

References

1. DDM & UNDP (2012) Annual disaster response, recovery and preparedness report. DDM, Dhaka
2. NIRAPAD (2007) Union disaster management system in Bangladesh. NIRAPAD Head Office, Dhaka
3. The Government of Bangladesh (2008) National disaster management plan. MoDMR, Dhaka
4. Islam T, Giri S (2011) Disaster management coordination in Bangladesh: effectiveness challenges. AusAid, Dhaka
5. BDPC (2–10) Disaster management coordination in Bangladesh. SDC, Dhaka
6. JICA (2007) Sustainable Physical Infrastructure and DRR in Bangladesh. JICA Office, Dhaka
7. SDC (2008) National coordination system on DRR. SDC, Dhaka
8. Khan M (2008) Disaster preparedness for sustainable development in Bangladesh. Disaster Prevention and Management Centre, Dhaka
9. UNDP (2011) Capacity development for DM coordinators. UNDP Bangladesh, Dhaka
10. The Government of Bangladesh (2008) The standing orders on disaster (revised). Ministry of Food and Disaster Management, Dhaka
11. MoDMR (2011) Coordination with donors—LCG for DER. MoDMR, Dhaka
12. ADB (2007) Disaster and development: why coordination is challenge for Bangladesh. ADB Dhaka office, Dhaka
13. Huq M (2008) Effectiveness of UDMC in disaster relief and recovery works. UNDP Bangladesh Office, Dhaka

14. Ali I, Hatta ZA, Azman A (2014) Transforming the local capacity on natural disaster risk reduction in Bangladeshi communities: a social work perspective. Asian Soc Work Policy Rev 8(1):34–42. https://doi.org/10.1111/aswp.2014.8.issue-1, https://doi.org/10.1111/aswp.12023

Chapter 5
Role of Bangladesh Government in Disaster Management

5.1 Introduction

It is important to understand the origins of the most devastating catastrophes that hit Bangladesh during the research period (2001–2012) and how they affected people's lives, livelihoods, and the economy before evaluating the government's efforts to mitigate risk. Located on the Bay of Bengal, Bangladesh has the world's longest stretch of uninterrupted sand, stretching for 120 km. Bangladesh is in a particularly precarious position due to its location, topography, abundance of rivers, and monsoon climate, all of which make it very susceptible to natural disasters. The severity of the effects of natural disasters in Bangladesh is affected by the region's distinctive coastline topography. Natural disasters, especially in the south and west, make coastal residents more vulnerable and delay economic and social progress.

Natural disasters such as floods, cyclones, droughts, tidal surges, tornadoes, earthquakes, river erosion, fire, infrastructure collapse, high arsenic levels in ground water, water logging, water and soil salinity, diseases, and other types of pollution are all very prevalent. There is now a greater potential for harm and vulnerability to communities as a result of climate change. Even though the scale of these shifts may seem small, they have the potential to greatly amplify the occurrence and severity of already-observed climatic disasters (floods, droughts, cyclones, etc.). Forecasts suggest that extreme weather events like floods and cyclones will become more often and intense and may even occur outside of their "known seasons." In certain locations, such as the Amazon, drought may previously have never happened, but this is no longer the case. Several natural disasters struck the nation between 2001 and 2012, with the floods of 2004 and 2007 as well as Cyclone Sidr and the cyclone of 2009 standing out as the most damaging.

Flood 2004

After earlier flooding in April had washed away most of the key annual rice crop in Bangladesh's northwest regions, the monsoon floods began on July 8. Northern areas along the Brahmaputra and Jamuna Rivers began experiencing flooding. Flooding

M. T. Islam, *Disaster, Governance and Development*, Disaster Studies and Management,
https://doi.org/10.1007/978-981-99-1631-3_5

affected much of Central Bangladesh and the Greater Dhaka area, notably areas near to the two confluence locations of the three powerful rivers and lasted for three to four weeks until finally emptying southward (Jamuna, Teesta and Bharmaputra). On July 24, the flood waters peaked at almost 35,000 km and spread out to their farthest. By mid-August, the water had subsided in most regions, but it lingered and sat still behind protective embankments where drainage was inadequate. There were a total of 36 million individuals impacted, or 25% of the population, throughout 39 out of 64 districts [1].

Midway through September of that year, a localized depression brought about six days of continuous torrential rain and high winds, which not only exacerbated the already-severe flooding in many parts of Central Bangladesh, but also flooded areas that had never been flooded by rivers before, such as Dhaka and other urban areas and some of the most productive agricultural land.

Figures from the government of Bangladesh (GoB) indicate that 36 million people (or 25% of the entire population) and 39 out of 64 districts were impacted. However, the official death toll was around 800; this is a testament to the efficiency of disaster response efforts and emergency health services. The total estimated cost of the damage from the 2004 flood is Tk. 11,418.6 crore. This sum was equal to 3.4% of Bangladesh's GDP in 2017 (Fiscal Year 2004) [2] (Fig. 5.1).

Homes, courtyards, kitchen outhouses, latrines, livestock buildings, haystacks, vegetable gardens, and paddy fields were all submerged, damaged, and eroded as a result of the floods.

Many residents were forced to relocate to higher, drier ground, a school, or the houses of friends and family, while others remained in their flooded homes to protect their belongings. Damage to farmland, fisheries, houses, educational institutions, roads, bridges, culverts, embankments, and flood-protection systems was discovered when the water receded in August. Road and rail transit, electricity supply, and telecommunications were affected in many areas of the nation.

Flood 2007

The Padma and Brahmaputra Rivers both flooded on August 1, 2007. 500,000 people were cut off from the rest of the nation as the primary route linking Dhaka to the rest of the country became impassable on August 3. Around 7.5 million people have already left their homes by August 7 [3]. There were over 50,000 cases of diarrhea and other watery infections by August 8, and over 400,000 people were staying in makeshift shelters. Approximately, 100,000 individuals in Bangladesh had had dysentery or diarrhea by August 11, and there had already been fatalities from the floods and an increase in the number of persons with flood-related ailments. In Bangladesh, the confirmed death toll reached 405 as of August 13. All six of Bangladesh's divisions were impacted by the disaster by August 15, and five million people had to be relocated as a result [4] (Table 5.1).

DMB data shows that paddy yields dropped by more than 40% in the flooded region. In every one of these towns, land productivity has dropped, impacting almost every kind of crop. There was a reported drop in rice output because to river erosion, inundation, floods, and sand deposition and in vegetable output because of moisture

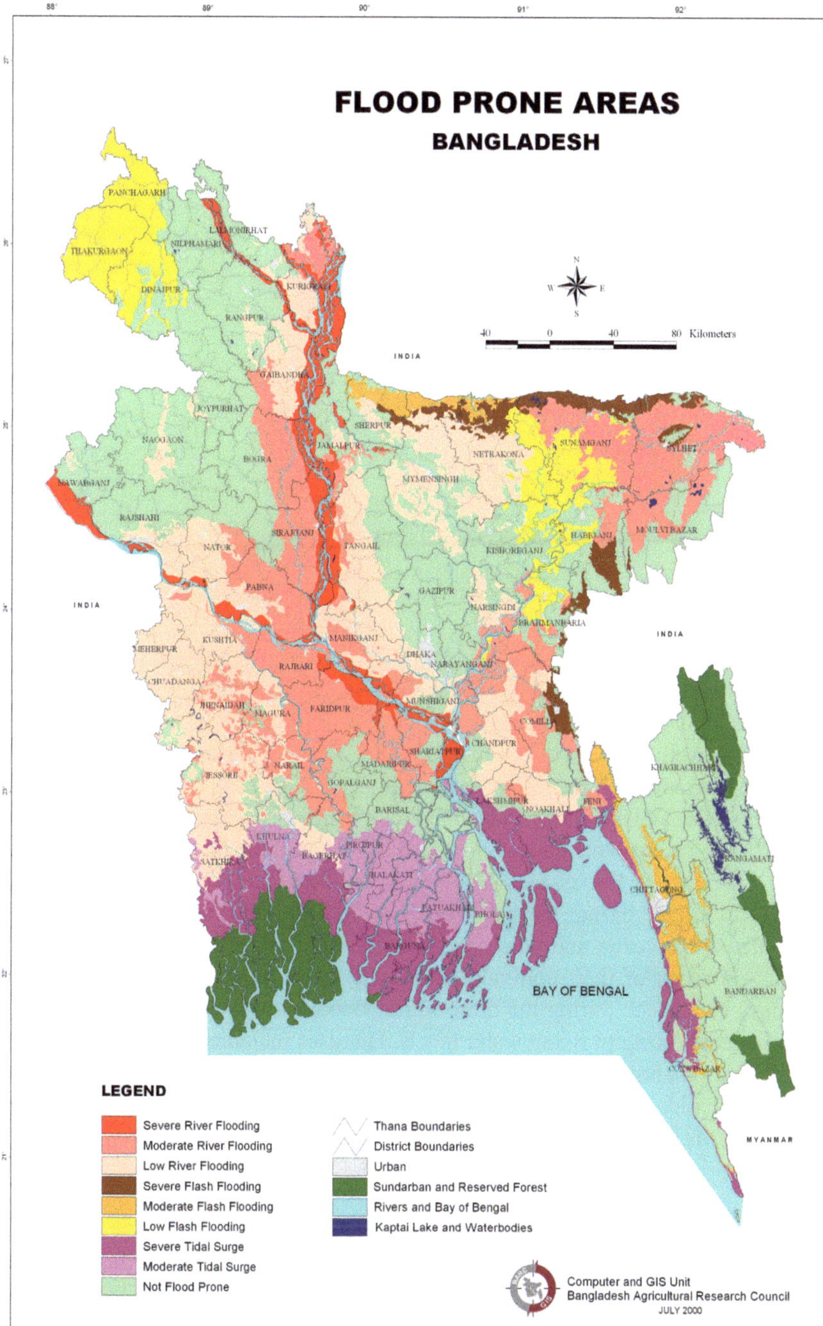

Fig. 5.1 Flood-prone areas in Bangladesh

Table 5.1 Key facts of Flood 2007

Affected population	10.5 million
Deaths	615
People affected by illnesses/injuries	68,862—Diarrhea 17,310—RTI 17,626—Skin diseases 4662—Eye infections 172—Snake bites 3284—Injuries 65,933—Other
Schools washed away by river erosion	44
Primary schools affected	4603
Primary schools closed	4444
Primary schools being used as flood shelters	292
Number of government shelters	401
Number of people in shelters	114,747

Source Disaster Management Bureau, 2007

and late cultivation. Flooding and inundation kept the soil moist, making it harder to plow for spreading winter crops, which in turn slowed down the production of rice.

Animals including goats and sheep, chicks, buffalo, and cows and oxen perished in the flood as well. It had an effect on the people's livelihood since animal husbandry is the second greatest source of revenue after agriculture, and because some cattle were swept away and many others died by flood from mysterious illnesses following the flood. The recent rains caused a food crisis for buffalo, while health problems plagued fowl. During floods, lack of animal food is a major problem in this region.

The local village bazaar and the weekly village market (known as Hat) are the only sources of revenue and sustenance for the residents of that char. The char's residents do much of their commercial activity in the neighborhood market. For many days each week during the monsoon or flood season, the village market had to be evacuated owing to rising water levels. Diseases including dysentery, cholera, jaundice, and skin issues were widespread because they were spread by contaminated water. Small children and the elderly were disproportionately affected by these health issues. Lack of convenient access to primary care services limited patients' options for therapy.

Cyclone Sidr 2007

On November 9, 2007, Cyclone Sidr was initially seen southeast of the Andaman Islands, with weak low-level circulation close to the Nicobar Islands. On November 11, while passing a little ways south of the Andaman Islands, it began to show signs of developing into a tropical cyclone, and by 13th of that month, it had become a cyclonic storm with hurricane-force winds concentrated around its center. Around 6:30 p.m. on the evening of November 15, Cyclone Sidr made landfall over the

Barisal coast during ebb tide. Sidr was a category 4 storm when it made landfall, with a circumference of about a thousand kilometers and sustained winds of up to 240 kilometers per hour [5] (Fig. 5.2; Table 5.2).

Nearly, 4000 people were murdered, but about 8.5 million women, men, and children were impacted. Nearly, 1.5 million dwellings were either totally or partly damaged. There were over 1.2 million animal losses and 2.4 million acres of crop failure. It was estimated that the economy had suffered a loss of $1.7 billion. The southwest coast of Bangladesh took the brunt of Cyclone Sidr's destruction. In terms of GDP, Cyclone Sidr had a little effect (GDP). Amounting to 2.8% of the country's GDP, it was a significant sum [6].

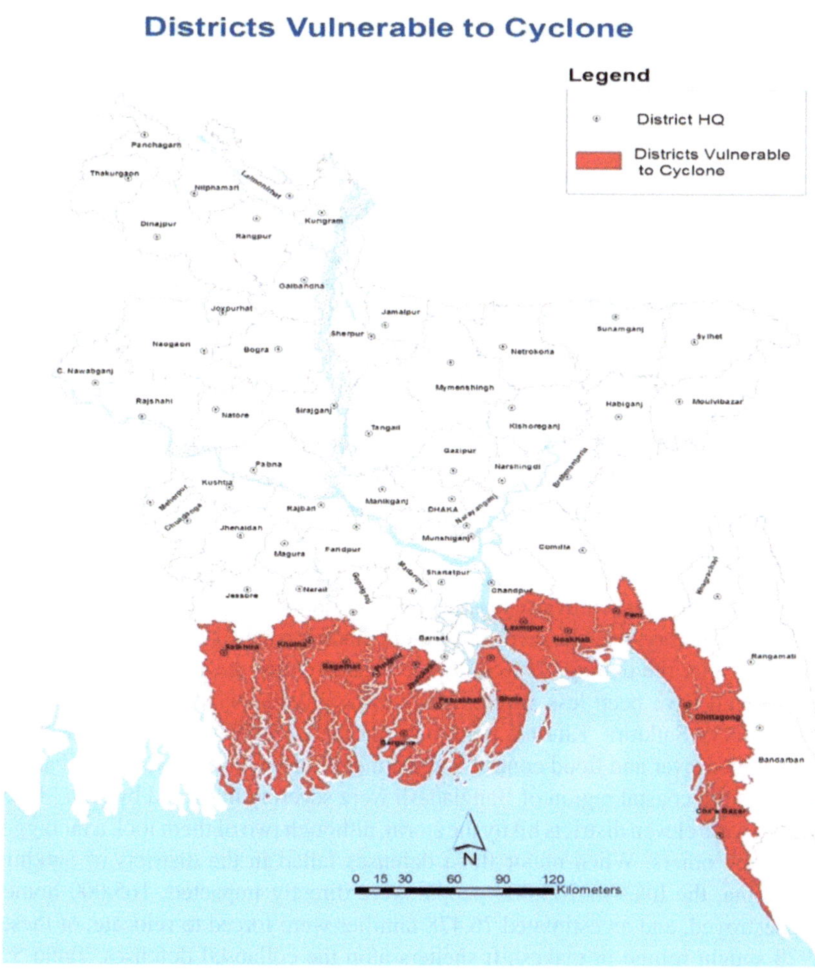

Fig. 5.2 Districts vulnerable to cyclone

Table 5.2 Impact of Cyclone Sidr at a glance

Affected families	2,064,026
Affected persons	8,923,259
People killed	3363
People injured	55,282
People missing	871
Damaged houses	1,522,077
Damage to crops (acres)	2,473,639
Damage to trees (number)	4,065,316
Total livestock dead	1,778,507
Educational Institutions damaged	16,954
Damage to roads, complete, and partial (km)	8075
Damage to bridges and culverts (number)	1687
Damaged embankments (km)	1875
Damage to electricity supply: 33 kV line and 11 kV line (km)	416 and 287

Source Disaster Management Bureau, 2007

Preliminary projections suggested that national economic growth would be damaged by less than 0.5% in the fiscal year of 2007–2008; however, much bigger drops were predicted to occur in the economies of the most afflicted regions. That decline was comparable to the predicted slowdown in growth after the 2007 floods, and it lowered the expected predisaster growth rate from 7% to 5% [7].

In the worst-affected areas, around 2 million people lost their jobs and income. Districts of Bagherat, Barguna, Patuakhali, Pirojpur, and Barisal were hit particularly hard, and they also had some of the worst poverty rates in the nation.

Cyclone Aila 2007

On May 25, 2009, the southwest coast of Bangladesh was battered by Tropical Storm "AILA," a category 1 cyclone. If the cyclone hadn't reached Bangladesh during a high tide cycle, the damage it would have caused (from the winds and the storm surge) would have been less severe. Many villages were wiped out, especially in the districts of Satkhira, Khulna, Bagerhat, Barguna, Patuakhali, and Bhola, as the waves eroded river and flood control embankments and dykes. Eleven total districts in the Western coastal region of Bangladesh were severely impacted by Aila [8].

There were eleven districts hit by the storm, although two of them took a far bigger hit than the others. When major flood defenses failed in the districts of Satkhira and Khulna, the lives of 255,000 people were directly impacted, 165,000 homes were destroyed, and an estimated 76,478 families were forced to relocate; of these, 25,928 sought refuge in makeshift shelters atop the collapsed defenses. Table 5.3 containing data on the destruction caused by Cyclone Aila, produced by the Disaster Management Information Centre (DMIC) of the Bangladesh Ministry of Food and Disaster Management.

Table 5.3 Key facts of Cyclone Aila 2007

Affected districts	11
Affected Upazilas	64
Affected unions	195 (fully), 334 (partial)
Affected households	948,621
Affected population	3,928,238
Reported deaths	190
People injured	7103
Houses damaged	243,191 (fully), 370,587 (partially)
Crops damaged	77,486 acres (fully), 245,968 acres (partially)
Livestock deaths	150,131
Roads damaged	2233 km (fully), 6621 km (Partially)
Educational institutions	445 (Fully), 4588 (Partially)
Embankment damaged	1742.53 km
Damaged bridges/culverts	157

Source Disaster Management Information Centre (DMIC), Ministry of Food and Disaster Management (2009)

With the main embankments having been breached at the beginning of the monsoon season, the cyclone's effects were magnified, and the Bangladesh Water Development Board (BWDB) being unable to fix the damage on its own, a humanitarian crisis arose that the government of Bangladesh could not solve on its own. It's fair to question if this disaster might have been foreseen and averted, given the Western Coastal Zone's preexisting vulnerabilities. The government of Bangladesh decided against designating Cyclone Aila a national catastrophe because an evaluation based on the danger factors and a simple comparison with the Cyclone Sidr phenomenon provided the incorrect signals. As a result, the reaction time of funders and NGOs was prolonged, and a more rapid and robust mobilization of all stakeholders was thwarted. Meanwhile, the situation continued longer than expected, necessitating gradual mobilization of the worldwide humanitarian community.

Here, we assess the efforts of Bangladesh's several administrations between 2001 and 2012 to lessen the impact of such catastrophes.

5.2 From 2001 to 2006: Strive Toward Professionalism in Disaster Management

For sustainable development to occur, it is necessary to lessen the negative effects of all natural hazards that have an effect on the socioeconomic situation. After coming to terms with this fact, the government of Bangladesh has implemented several policies and initiatives aimed at catastrophe mitigation. Since 1993, when the BNP-led government of Bangladesh recognized the importance of disaster management in relation to risk management, they've been working on a project called "Support to Comprehensive Disaster Management," whose overarching goal is to lessen the devastating effects that natural disasters have on the country's people, economy, and environment [9].

The project's development aim included helping families and communities in disaster-prone regions create Local Disaster Action Plans (LDAPs) to better prepare for and respond to natural disasters like cyclones and floods. The development aim of the initiative included training and creating awareness as a key component. When this project was finally wrapped up on June 30, 2001, it cleared the way for a more comprehensive approach to risk management to be developed with the help of development partners and international organizations under the Comprehensive Disaster Management Programme (CDMP). In the backdrop of reducing poverty and ensuring Bangladesh's continued progress, Begum Khaleda Zia's administration made CDMP a top priority.

To better coordinate the disaster management system as a whole, the government of Bangladesh prioritized both structural and non-structural mitigation measures in light of the new disaster management concept. The government of Bangladesh is firm in its belief that structural mitigation measures are necessary to alter or decrease certain catastrophe consequences, in addition to non-structural mitigation measures. In order to reduce the likelihood of disasters, initiatives in Bangladesh prioritize both structural and non-structural measures.

The nation was hit by devastating floods in 2004 when the BNP was in power. The BNP government first believed that the media was exaggerating the severity of the damage caused by the 2004 floods. Fast-paced developments need government attention, especially given the unexpected (flash) and devastating character of Flood 2004. The flood waters began to rise at the beginning of July, and the government's relief efforts kicked off during the second week of the same month. A week later, on July 22, 2004, the National Disaster Management Council convened. Since Bangladesh had "sufficient supply of foods," the committee determined that the country would not go outside for aid. The council did vote, though, to seek donor aid for post-flood restoration. Additionally, it was determined that Flood 2004 would not be considered a national emergency. It took some convincing for the administration to allow NGOs to participate in the relief effort.

Almost four weeks after the flood began, on July 26, 2004, government officials finally met with the NGOs (a total of fifty-six). The government has requested NGOs to "inform the government" about their relief-related operations and has encouraged them to launch aid programs. In contrast to 1998, when the government first called for coordinated relief and housing efforts, neither initiative was coordinated. Furthermore, the government's perspective on overseas aid went through three distinct phases:

- Assistance not required for relief (July 22, 2004).
- Assistance required for post-flood rehabilitation (July 27, 2004).

- Call for comprehensive international assistance (July 28, 2004).

Government headed by Prime Minister Begum Khaleda Zia has been cautious for a number of reasons. One was the nature of the flood itself, as water levels rose surprisingly quickly in several places. The government would have been able to evaluate the situation with the help of an early warning system. However, this wasn't there. The government's confidence that flood damage would be minimal and that it could handle the situation with its own food stocks and domestic efforts may have played a role in its decision not to declare the Flood of 2004 a national emergency and not to seek food help [10]. It's possible that the administration's desire to "project" domestically and internationally as a government in charge was tied to the showing of its capacity to handle current problems. Declaring a national catastrophe or asking for foreign aid may have been seen as bad for the economy in the near term since it may discourage buyers and investors from doing business with Bangladesh.

As an additional vote of confidence, the 2004 floods occurred immediately after the announcement of the fiscal year's budget, which left a sizable quantity of money in several sectors (under titles such bloc allocation, repair and maintenance, etc.) that could be transferred for relief and rehabilitation. Since flood waters began to diminish in most areas by the end of July, the decision to not proclaim a state of emergency may have been appropriate. As a result of mounting concerns of a catastrophic flood, the government of Bangladesh (GoB) under Prime Minister Begum Khaleda Zia implemented its Standing Orders on Disaster. There had been an initial round of meetings of the National Disaster Management Council, which was chaired by Prime Minister and tasked with assessing the flood situation and deciding on the national response [11].

The government of Bangladesh has set up an Emergency Operation Centre in the nation's capital to collect and analyze data from all of the country's individual districts. The administration has also continued to welcome foreign aid, and it has urged the United Nations and the NGO community to back the country's relief and, in particular, its recovery and rehabilitation efforts. Despite this, foreign players contributed to the reaction and recovery, supplementing the efforts of the government.

Various UN organizations have sent much-needed items to the administration, including water purification pills, antibiotics, and bleaching powder. Due to poor communication, many Upazilas ended up responding to the same situation. Numerous reports of favoritism, theft, and politicization have surfaced.

The District, Thana, and Union all formed disaster management committees. In actuality, there was extremely inadequate coordination to minimize duplication of efforts, and NGOs did not coordinate with one another at the district or Upazila levels. The only purpose of the Disaster Management Coordination Committee meeting was to monitor government relief efforts. Coordination committees failed to oversee the government's, NGOs', and donors' long-term rehabilitation activities. There was not a consistent schedule for the NGOs Coordination Meeting. The rehabilitation effort was threatened by ineffective local leadership. Politicians were not keen on coordinating closely with city hall to speed up the rehabilitation process after the disaster.

For those who could be displaced by a cyclone or flood, the BNP government had built 1841 cyclone shelters and 200 flood shelters using both internal and foreign funds as part of its structural efforts for greater disaster preparation. Additionally, 482 water and flood control projects of varying sizes have been completed and drainage systems and a lengthy flood wall of 3400 kilometers in length as well. As a precaution against tidal waves, storm surges, and floods, 9000 sluice gates and regulators were built along various rivers and canals. The BNP administration took a number of important efforts toward establishing national and union-level institutional frameworks for efficient and methodical disaster management, which has helped to lessen the suffering of those who have been affected by natural disasters in Bangladesh. The BNP government had formulated a set of mechanisms for council and committees from national down to the grass-roots levels to maintain proper coordination among the concerned ministries, departments, line agencies, Local Government Body (LGD), and community people, and to ensure their proper functioning to mitigate sufferings of the people.

We found that the BNP regime prioritized a relief-centric approach to disaster management at the expense of catastrophe risk reduction, one of the important stages of the Disaster Management Model. There were no laws passed at the time that would have helped spread the word about DRR's value. At both the national and union levels, the committee structure was only an idea. Relief efforts were the only focus of disaster management. Relief efforts were prioritized since the BNP administration lacked a long-term strategy for disaster management.

5.3 From October 2006 to January 2009: Momentum Started for Promoting DRR

In 2007, on November 15, hurricane Sidr ripped across the southwestern part of Bangladesh, causing widespread destruction under the Caretaker Government. The storm slammed Bangladesh on November 15, killing at least 3268 people and wiping out whole communities along with their crops, infrastructure, and water supply. The interim administration of Bangladesh has made a plea for $1 billion in assistance to help repair the country's south coast after it was destroyed by a hurricane last month [12]. The military-backed caretaker administration claims the funds are necessary to rebuild after the storms and improve cyclone defenses.

Mr. Fakhruddin Ahmed, Head of the Caretaker Government, also stressed the necessity, stating that after the devastating floods brought on by the monsoon this year, the nation needs to adopt more preventative measures. Having endured two monsoon floods in a row and then Cyclone Sidr, we need renewed determination to control our future. There has been an increase in the intensity and frequency of natural catastrophes in Bangladesh as a result of climate change, and the chief of the caretaker administration has encouraged the international community to adopt a long-term, comprehensive strategy to supporting the country. In order to minimize the loss of cyclone Sidr [13];

- $250 m was needed to rebuild coast banks.
- $150 m was needed for reforestation.
- $300 m was needed to repair roads and bridges.
- $100 m was needed for damaged schools.
- $200 m was required for new cyclone shelters.

Cyclone Sidr ravaged numerous coastal areas, causing extensive destruction. Nearly, 4000 people were murdered, but about 8.5 million women, men, and children were impacted. Nearly, 1.5 million dwellings were either totally or partly damaged. There were over 1.2 million animal losses and 2.4 million acres of crop failure. We estimate total economic damage to be about $1.7 billion [14]. Efforts were made to mobilize a massive humanitarian response by the caretaker administration, military personnel, and innumerable volunteers, as well as national and international NGOs. Numerous foreign countries and organizations that provided aid responded quickly and generously. Local non-governmental organizations generally provided aid inside their own communities.

The government often acted quickly to begin providing aid. Many more showed up after we did. This does not prove that the government acted promptly, however. In certain locations, local elites and pro-Caretaker Government figures were discovered to have taken the initiative to provide aid for Cyclone Sidr victims on their own. Since distribution authorities were aiming to help the greatest number of people possible, the quantity of government aid in certain places was discovered to be far lower than necessary.

Help for rural communities was often inadequate because of government inaction. Disaster victims in Sidr thought that aid operations were mismanaged. The amount of food, cash, and clothes provided by the government for cyclone victims was inadequate. No one was discovered to have obtained all of the life-sustaining necessities that were promised. Some were fed just saline while they were starving, while others were given rice and pulses but no seasonings like salt, oil, or onion. Most households lacked the necessary utensils for cooking. Those under the age of two were found to be severely lacking in infant nutrition. As far as anybody could tell, there was no distribution of infant food aid.

The local health centers were found to provide sporadic and inadequate health services. Due to a lack of medication and saline, they were unable to offer proper care to the impacted population. In several of the communities surveyed, the distribution of aid was haphazard and insufficient. As opposed to those who had sought cover along the highways or on the embankments, those who had been impacted in more rural locations had limited access to the assistance effort. Those who have sought refuge in emergency flood shelters are receiving the most of the government's aid. Those who chose to live without safety were not receiving sufficient aid from the government.

The Upazila government lacks the resources necessary to provide aid by sea or air. U.N. officials and others made valiant efforts to aid as many storm victims as possible, but logistics proved difficult. There were discrepancies in the way the administration handed out aid. As a rule, the local chairman's staff compiled the list of impacted individuals without ever speaking to them directly. The majority of those surveyed have negative views about several political officials. Because these lists were hastily compiled, many of the impacted individuals were left off.

The Caretaker Government at the time was under pressure to manage the main political parties despite having the backing of the Bangladeshi Army. Both of the country's major political parties were told loud and clear that the military-backed administration needed new leadership. The government attempted to remove both Sheikh Hasina and her archrival, Bangladesh Nationalist Party Prime Minister Khaleda Zia, from office in early 2007. The administration attempted to expatriate Ms. Hasina in April by preventing her from returning from a trip to the United States. It also made an effort to get Ms. Zia to voluntarily depart the country. However, the government eventually caved, allowing Khaleda Zia to stay in the country, and Sheikh Hasina to return to a hero's welcome from inside her own party. They therefore devoted less attention to managing the aftermath of major disasters. They were only in office temporarily, so they didn't give much thought to how to make DRR a part of the legislative process. They kept in touch with the worldwide community to solicit financial and logistical support in order to help catastrophe victims rebuild their lives.

5.4 From January 2009 to the Date: Paradigm Shift in Disaster for Development

When Awami League President Sheikh Hasina took office as Prime Minister on January 6, 2009, the caretaker administration officially ended. Bangladesh Awami League's new path as governing party was put to the test on May 25, 2009, when Cyclone Aila ravaged the country's coastal districts. Sheikh Hasina's administration did an excellent job of dealing with the storm, and as a consequence, fewer people perished and their homes and livelihoods were not destroyed [15]. As a result of Cyclone Aila's relatively low intensity and its longer-than-anticipated effect, the worldwide humanitarian community responded in phases and had to continually adjust to changing conditions on the ground. The international community's reaction was appropriate to the size of the catastrophe, reaching a total of 84.7 M US$, but stretched over 3 years in four subsequent phases. This was in addition to the response and recovery plan launched and conducted by the Ministry of Disaster Management.

Long-term rehabilitation efforts in regions hit by Cyclone Sidr were also a priority for the Awami League Government as they dealt with Cyclone Aila. Based on our observations, political leadership and local administration are both present as victims of Cyclone Sidr begin to rebuild their lives. The long-term recovery plan for storm

Sidr victims was reinforced by the Disaster Risk Reduction Programme [16]. In 2008, USD 85 million were used to help those who had lost everything in the storm Sidr rebuild their homes and livelihoods.

Prime Minister Sheikh Hasina successfully led the government to implement a $30 million Aila Recovery Programme. The Prime Minister has ordered the District and Upazila Administration to convene a Disaster Management Coordination Committee twice a month, and he has instructed concerned Members of Parliament to monitor the relief and restoration efforts.

It was because of strong political leadership that the reaction and recovery operations were carried out effectively. Donor organizations from across the world provided USD 50 million to supplement the Awami League Government's efforts to aid in the long-term rehabilitation of coastal districts hit by Cyclone Aila and to spread awareness of disaster preparedness and response. The United Nations, in cooperation with the regional Member of Parliament, established a twice-monthly meeting to ensure coordination among NGOs [17].

After taking government in 2009, the Bangladesh Awami League reevaluated the goals of the previous administration's Comprehensive Disaster Management Project (CDMP) Phase II in an effort to raise awareness of the need of reducing disaster risk in the country. For the first time in Bangladesh's history, the political administration led by Prime Minister Sheikh Hasina has mainstreamed the idea of DRR via the implementation of legislation and the reorganization of institutions at all levels, from the national to the local. For example, in 2010, the Awami League showed their deepest degree of dedication to DRR by revising the SODs and endorsing the National Plan for Disaster Management. Next came the passing of the National Disaster Management Act in 2012, which was a game-changer for the country's preparedness for and response to disasters.

The new legal framework and altered institutional structures have allowed Bangladesh to reflect on catastrophes and implement the paradigm shift from relief and rehabilitation to disaster risk reduction (DRR) in all important policy papers. Also, this new area has emerged at a time when initiatives are being launched to integrate climate change adaptation and disaster risk reduction and when the identification and comprehension of new and emerging risks are a source of anxiety. When it comes to international catastrophe management, Bangladesh's policy orientation is still cutting edge. Bangladesh is already adopting the present international policy formulation, which emphasizes integrating the post-2015 agenda with the post-HFA agenda and the convergence of development and risk reduction policy. Now, this theory has to be put into action in the field of development.

The real actors in emergency management, including search and rescue, distributing reliefs like food, drinking water, and medicine, and disseminating warning messages are members of volunteer organizations like scouts and girls guides as well as members of community-based organizations and government bodies like Ansar and Fire Service and Civil Defence [18]. The government of Bangladesh is actively working to improve its capability for risk mitigation and disaster response management via training and awareness raising workshops.

The current government in Bangladesh helped scouts establish a specialized Disaster Response Management Unit, which trains hundreds of scouts and scout professionals each year to respond more effectively to disasters. In addition, the people who volunteer for the Cyclone Preparedness Plan (CPP) in Bangladesh's 13 coastal districts have been given a structured capacity-building programme to help them spread an early warning message to the local population. In order to reduce the underlying risk connected with disasters, the current government of Bangladesh has placed an emphasis on training to aid concerned authorities and volunteers in carrying out a DRR plan.

Using information and communication technologies to better handle emergencies is a primary priority for the current Awami League administration. Mobile technology that employs Short Message Service (SMS) and Interactive Voice Response (IVR) to disseminate weather advisories and catastrophe early warnings is now available to all mobile carriers worldwide. As a consequence of CDMP II's lobbying, the Ministry of Education and the Ministry of Primary and Mass Education issued a government order mandating regular school safety exercises (at least twice a year). At response to this directive, earthquake exercises were carried out in 30,000 elementary schools and 12,000 secondary institutions throughout the country.

The government has established ambitious targets for disaster risk reduction and emergency management by 2015, as outlined in the National Plan for Disaster Management 2010–15. As a result, the government's commitment to achieving these objectives and plans for disaster risk reduction and emergency response should have been made clear in the national budget for 2013–14. The Awami League Country's Minister of Finance estimated that between 1990 and 2008, natural catastrophes cost the government $2,189,000,000, or around 1.8% of GDP. The Ministry of Disaster Management and Relief (MoDMR) has been given a total of Tk.6524 crore (development and non-development budgets combined) for the 2013–2014 fiscal year. In addition, the SSNPs received Tk.25,371 crore (11.4% of the total budget) in the FY 2013–14 budget [19]. The MoDMR received a substantial portion of this budget's allotment. Ultimately, the most crucial factor for MoDMR to think about is how this SSNP may transition into social security programs that will safeguard vulnerable individuals.

Beneficiaries of SSN programs should be selected with regional susceptibility to natural disasters and poverty in mind. The government is building 74 flood shelters in flood-prone and river-erosion zones and 100 cyclone shelters in coastal locations to protect residents from natural disasters. In addition, a total of Tk.69 crore's worth of equipment has been acquired to help with earthquake and other catastrophe preparation. The increasing abnormality of various natural disasters is raising fundamental questions about the role of government, how it would manage the disaster, and what would be the more rational and effective way of management in light of the emergence of the changing roles of the government and the increasing participation of the public in the government's sociopolitical activities.

Policymakers, experts, and professionals are coming to agree that the government, working alone, cannot and will not adequately manage and handle all sorts of catastrophes with its equipment. Both industrialized and developing nations have been urged to adopt this attitude, which emphasizes the need of enlisting the help of local people, leaders, and communities to offer disaster relief during and after an event. In light of this attitude, a new disaster management strategy has emerged: the community-based approach (CBA), which places an emphasis on the involvement of every member of the community in response to an emergency.

The current Awami League Government in Bangladesh is steadily spreading support for a Community-based Disaster Management approach. The idea of Community-based Disaster Management was included into the National Disaster Management Plan in 2010. Local residents are encouraged to participate in disaster relief, reconstruction, and risk mitigation efforts by a provision in the National Disaster Management Plan [20]. The Department of Catastrophe Management has asked the Ministry of Disaster Management and Relief for advice on how to include locals in disaster preparation and response from the outset. One example of community-based disaster management is the Project Implementation Committee. Additionally, the union disaster management committee will include community members.

The Awami League Government's successes include the introduction of the Citizen Charter in Disaster Management and the subsequent amendment of the Disaster Management Standing Order. Quality and quantity of government disaster response, recovery, and risk reduction services are agreed upon in the Disaster Management Citizen Charter, which serves as a contract between the government and its citizens. It is a detailed description of the services to be provided by the organization, the fees to be collected from the people receiving those services, the time it will take to provide those services, and the channels through which people can file complaints if they are denied those services, all of which were developed in collaboration with community members.

The Awami League Government, under by Prime Minister Sheikh Hasina, has made significant strides toward its goal of making DRR the norm in disaster management by reducing the emphasis on aid.

5.5 Paradigm Shift from Relief-Centric Approach to Disaster Risk Reduction

Over the last decade, the term "paradigm shift" has grown more important in Bangladesh's disaster management lexicon. The Awami League government of Bangladesh, led by Prime Minister Sheikh Hasina, inspired the country to believe that a paradigm change inviting DRR as the major business of disaster management was possible. In the disaster management context of Bangladesh, the term "paradigm shift" is shorthand for the transition from emergency response and rebuilding to preparedness and prevention. The disaster management community in Bangladesh has a firm grasp on the paradigm shift idea, and its essence has been incorporated into the national policy discourse. The Comprehensive Disaster Management Programme is the product of this idea in programmatic form (CDMP). The Ministry of Disaster Management and Relief (MoDMR) and the Department of Disaster Management (DDM) are working together on a UNDP-funded initiative to strengthen risk management procedures at all levels of government. Numerous indicators suggest that the paradigm change has also been welcomed by NGOs and funders.

The most prevalent measure of the paradigm shift's effectiveness is a decrease in fatalities caused by disasters. Though this is cause for celebration, it does not imply that Bangladesh is immune to natural catastrophes or that the country's disaster response infrastructure does not still need additional development. CDMP is a risk reduction initiative that has prioritized the adoption of the Hyogo Framework for Action and the incorporation of DRR into all levels of government. Disaster recovery is not a focus of CDMP's activities. It has been widely noted that the disaster management committees' (DMCs) primary ongoing activities continue to be reaction and relief. The most up-to-date strategy is making communities more resilient so that they can better confront and manage natural disasters on their own. There is a strong emphasis on the use of local knowledge and indigenous resources to strengthen communities.

5.6 Recurrent Disasters as a Consequence of Climate Change: Climate Change Diplomacy—Best Alternative to Adaptation

Increasing evidence of both the consequences of and human contributions to climate change make it one of the most important concerns of the twenty-first century. Worries are rising in many parts of the globe that climate change will have to be placed within a larger foreign policy framework in order to garner the required agreement and commitment to action. Stability and security, as provided by traditional foreign policy, are seen as essential to human flourishing, international peace and freedom, and economic development across the world. Traditional diplomatic tools are not always helpful in dealing with global challenges in today's more linked globe.

When greenhouse gas emission is not the goal of any one "unreceptive" influence, it might be difficult for established alliances and processes to be successful against a problem like climate change. Foreign policy needs fresh thinking to face the problem of climate change, thinking that takes into account climate change involvement beyond the confines of the environment. This "marriage of convenience" between climate scientists and diplomats and foreign ministry officials is inevitable in the field of climate change diplomacy, despite the fact that these two groups do not often share a same language.

The principle of "common but differentiated responsibility" provides the best paradigm and institutional framework to understand and confront the asymmetries in the international system and is widely agreed upon by experts in both science and politics as the best way to address the unique needs of vulnerable countries like

Bangladesh and the Maldives. Bangladesh is one of the nations most at danger from the predicted effects of climate change despite the fact that it is not a major contributor to the problem. Cyclones, tornadoes, and floods, all examples of extreme weather, have grown increasingly common and destructive in recent years. Bangladesh is in a precarious position due to its geographical position, topography, abundance of rivers, and monsoon climate, all of which increase its susceptibility to natural disasters. Hazards in Bangladesh are affected by the region's unique coastline topography. Natural disasters, especially along the southeast coast, make coastal residents more vulnerable and impede economic and social progress.

The world is becoming more and more conscious that addressing climate change requires putting it in the context of foreign policy as a whole in order to get the support and dedication needed to finally take action. The stated purpose of Western foreign policy is to ensure peace and safety so that all people everywhere may enjoy the fruits of democracy and economic growth. Traditional diplomatic tools are not always helpful in dealing with global challenges in today's more linked globe. The source of climate change (greenhouse gas emissions) is not the ambition of any particular "enemy" power, making it difficult to respond to with the same alliances and methods that have been used in the past.

Foreign policy has to be rethought in light of the climate change problem, rethought in a way that takes into account climate change involvement not only in the environment sphere, but also outside the environment box. The foreign policy of the country is moderate, and it relies heavily on international diplomacy, particularly in the United Nations. Since gaining independence in 1971, the nation has governed by the idea of "friendliness for everyone, enmity against none" in its diplomatic relations. Bangladesh is a non-aligned member state and hence does not officially support any one of the world's main powers. The government has been working to improve ties with its neighbors in the area since the conclusion of the cold war. Bangladesh's foreign policy is predicated on the belief that every people has the inherent right to freely select and construct its own social, economic, and political system via means of its own free choice.

In doing so, Bangladesh is showing its commitment to safeguarding the human rights of those in the country who are vulnerable to the effects of climate change. To solve these basic issues, Bangladesh is increasingly seeing negotiation at all levels as a necessary and useful activity. The diplomatic approach to climate change emphasizes both prevention of the worst impacts and preparation for them. Bangladesh, as an example, prioritized the following three measures during the Durban negotiations: (1) create financial sources based on international commerce; (2) specify yearly objectives for scale-up; and (3) adopt a transparent, centralized accounting system.

The government of Bangladesh's Ministry of Foreign Affairs (MoFA) plays an important part in international diplomacy over climate change. The Ministry of Foreign Affairs (MoFA) is charged with carrying out this duty on the global stage, either via bilateral or multilateral means. There is evidence that Bangladesh is playing a significant role in this arena. An outstanding example of Bangladesh's achievement in climate change diplomacy is the 2011 ministerial level conference of the Climate

Vulnerable Forum (CVF) held in Bangladesh jointly coordinated by the Ministry of Environment and Forest (MoEF) and the MoEF.

The CVF meeting was able to reach a consensus to raise a unified voice for the interest of vulnerable states at a separate summit since the UN Secretary General was there. To hasten talks on this issue, Bangladesh has ramped up its efforts to combat climate change via the creation of the Bangladesh Climate Change Strategy and Action Plan (BCCSAP). The BCCSAP places a premium on productive negotiation engaging expert envoy(s) at the national and internal level to address the worst effects of climate change. By actively participating in a variety of climate conferences, such as COP, Bangladesh showed that it can serve as a global leader in gaining the benefits of productive discussions (Conference of Parties).

The goal is to evaluate Bangladesh's participation in the negotiations at each level so that more aid may be allocated to the country to help it adapt to climate change. Content analysis is the focus of this study. However, it included a comprehensive review of pertinent publications, books, and other materials and took place in Dhaka, Bangladesh, between July and September 2012. The research began by looking at how the government of Bangladesh's Ministry of Foreign Affairs participated in its multi-tiered climate change negotiation efforts. Consecutive efforts are being made to investigate climate change's impact on Bangladesh.

When it comes to the effects of global warming, Bangladesh is among the most at-risk nations. Current scientific knowledge predicts that climate change will pose a significant danger to the health and survival of the people of Bangladesh over the next few decades.

As the most vulnerable nation on earth, Bangladesh is feeling the effects of climate change. Predictions on the changing pattern of temperature, rainfall, and sea level rise in Bangladesh as a result of climate change were emphasized in the National Adaptation Programme of Action (NAPA) (Table 5.4).

Based on these climatic parameter projections, we can say that the country will be extremely vulnerable to (a) increased flooding, both in terms of extent and frequency; (b) increased moisture stress during dry periods leading to increased drought susceptibility, both in terms of intensity and frequency; and (c) increased salinity intrusion during the low flow conditions. There are many major productive systems that will be directly impacted by these shifts in the physical system of the country, including (a) crop agriculture, (b) livestock production, (c) aquaculture and fish production,

Table 5.4 Statistical data of predication for Bangladesh due to climate change impact

Year	Temperature change (°C) mean	Rainfall change (%) mean	Sea level rise (cm)
2030	1.0	5	14
2050	1.4	6	32
2100	2.4	10	88

Source National Adaptation Programme for Action, Department of Environment, Government of Bangladesh, 2005

(d) coastal shrimp production, (e) forest and vegetation, and (f) the livelihoods of poor and marginal households. Temperature and humidity changes will have knock-on effects on human health. Settlement patterns and permanent structures will be impacted by the high vulnerability to water-based natural disasters. The following parts, based on secondary sources, provide a concise explanation of how climate change is expected to affect the country's bio-physical features.

It is clear that Bangladesh is making an effort to maximize the benefits of its diplomatic connections in order to tackle the problem of climate change. For instance, China recognizes and appreciates Bangladesh's worries over this issue. China has always backed Bangladesh and other LDCs in their genuine and acceptable climate change objectives throughout international discussions. It is imperative that China, India, Malaysia, and Bangladesh, all of which are considered developing nations, adhere to the idea of "shared but differentiated duties" and cooperate to protect the interests of all developing countries. The effects of global warming are being seen in both China and Bangladesh. Regarding this matter, the problems that each country faces are same, as are the fundamental interests that unite us.

At the same time, Bangladesh has made an effort to work with China on the problem of climate change, since China has previously performed adaptation cooperation with Bangladesh. Bangladesh, for instance, received help from China with river dredging operations. The government of China sponsored training for Bangladeshi authorities and technicians. In order to promote "equal consultation, mutual benefit, and shared progress," Bangladesh should work to expand bilateral collaboration in this area.

The government of Bangladesh presents itself in two distinct ways: as a "worst victim," "peace-loving," and "responsible actor," and as a "poor developing nation," all of which are used to further the country's diplomatic ambitions. While the language of a "poor developing country" highlights the need for foreign help to address the negative consequences of climate change, the vocabulary of a "responsible player" enables Bangladesh to improve its international image and pursue its goals more readily.

In reality, talks provide the impetus for resolving a worldwide issue. There is a need for global cooperation for this reason. What is more, the actual bargaining goes down in the anterooms, ante-pool lounges, and other off-limits areas of the hotel. Since the turn of the twenty-first century, however, Bangladesh has taken a more active position in a number of global discussions aimed at addressing the problem of climate change, thanks to the lessons it has learnt over the years. Bangladeshi representatives to the United Nations framework convention on climate change were successful in their negotiations because they presented the country's true case as the worst victim nation. It was determined that Bangladesh made significant strides in taking a leadership role at the last COP. First, it's important to take an objective look at what role Bangladesh has played so far so that we can figure out how to build on this success and make it even better in the future.

In the recent climate change summit in Doha, Bangladesh played a pivotal role (November 2012). Even though they are not to fault for global warming, climate-vulnerable nations like Bangladesh have made strong requests for compensation. To show its future commitment to climate discussions, Bangladesh served as CVF chair, a position that required it to negotiate with wealthier nations to cut carbon emissions in order to maintain their obligations.

At the summit, poor nations demanded that wealthy nations lay out a plan for allocating the Green Climate Fund's projected $100 billion in resources between 2013 and 2020. The concerns addressed by the long-term cooperative action are ones that will not be resolved until far beyond 2020, such as the execution of the objectives that should be included in new climate change treaties and the fulfillment of the promises made by different nations to reduce their emissions. The Durban platform for expanded action and subsequent accords should further implement long-term targets for reducing emissions and negotiation outcomes depending on the action. While everyone involved seemed to be on the same page, it's clear that they know it won't be simple to accomplish their goal in the way it was envisioned at the outset. In order to achieve this goal, it is recommended that all parties to the contract participate in open dialogue and talks.

Until now, we may say that Bangladesh has played a "prominent" role, but by no means a "leading" one. They work hard to bring attention to Bangladesh's status as a climate change sensitive nation and the steps the government is doing to address this issue. However, it is believed that Bangladesh is merely promoting itself, and this is not seen favorably by its partners within the LDC group or other vulnerable nations. Bangladesh must gain the respect of its peers before it can assume leadership roles in international organizations. Regrettably, it undermined its own credibility by boasting about itself. Therefore, the country has to be able to take the lead in international forums like the Council on Foreign Relations (COP) if it wants to be taken seriously.

Bangladesh should think about designating high-level climate change envoys to negotiate on the country's behalf on climate change. Investing in a high-level special envoy with the government's confidence and the requisite diplomatic (rather than technical) experience and abilities, together with a team from key ministries, is crucial even today since climate change will remain a priority in Bangladesh for many years to come. In doing so, they will be able to take the moral high ground and pressure other nations (both developed and huge emerging countries that have become significant polluters) to take action. Instead of making an a priori claim for a certain proportion of any additional international adaptation money, Bangladesh should instead demonstrate that it deserves to get its fair part based on its performance and capacity to do the right thing with transparency and good governance.

If Bangladesh can prove it can responsibly handle financial resources, it will automatically be given more aid than it needs. Since addressing climate change is crucial, Bangladesh should employ a senior or retired ambassador with the requisite diplomatic skills and expertise as its permanent climate change negotiator. There is a general consensus that Bangladesh is one of the most vulnerable nations. We must present ourselves well in every region of the globe. In contrast, Bangladesh was ahead of the curve when it came to creating a climate strategy, climate action plan, and climate trust fund. Without waiting for contributions from other nations, Bangladesh has set up the fund immediately. Now that the spotlight has shifted to Bangladesh, the task will be to put the money to good use. If the funds are mismanaged, Bangladesh's image would suffer. The fact that weak nations are demanding aid is problematic. Of course we need to allocate more money. But requesting funds alone isn't enough.

If the business sector in Bangladesh is serious about alleviating human suffering on a regional, national, and international scale, it will need to step up. Companies in the energy, transportation, industrial, and other sectors have been identified as the primary sources of pollution on a worldwide scale, rather than governments. There has to be a shift in the way firms are managed in the private sector. They must switch to renewable power. While Bangladesh's environmental protection laws are strong in theory, their enforcement has been spotty at best. The government's own employees break the law on occasion. The laws of Bangladesh need to be strictly enforced. Industry leaders in Bangladesh, as well as ordinary individuals and the government who are in the best position to identify problems, need to work together more effectively. When it comes to international negotiations, Bangladesh is in a situation somewhat unlike that of the least-developed nations.

Bangladesh is particularly sensitive to the effects of climate change while being among the countries that emit the fewest greenhouse gases. Baseline nations like Brazil, South Africa, India, and China produce a disproportionately large share of global emissions due to their advanced economies and rapid rates of economic growth. Bangladesh must convince both wealthy and developing nations that although they have the right to progress economically, they do not have the right to destroy the environment in the process. Bangladesh must convince them that technological advancements may be made while maintaining low pollution levels [21]. Two main tenets of Bangladesh's foreign policy are ensuring security and fostering economic growth. Security, commerce, exporting of human capital, FDI, philanthropy, international assistance, cultural concerns, preventing the spread of terrorist ideologies, and protecting the environment are just few of the many topics included by foreign policy.

Diplomacy on climate change plays a crucial role in achieving agreement and mobilizing resources for dealing with the effects of climate change. The following consequence (adaptation and mitigation to climate change) of climate change diplomacy is shown in Fig. 5.3, a liner flowchart created by the author himself.

There are intricate connections between international and national politics that characterize climate change diplomacy. New players emerge as a result of the environmental diplomacy's intrinsic relationship between international and domestic policy. Cooperation and healthy rivalry for worldwide environmental leadership are both essential for effective environmental diplomacy. Previously distinct spheres of foreign policy, such as trade and investment, development and human rights, and even military security, are increasingly linked with environmental issues.

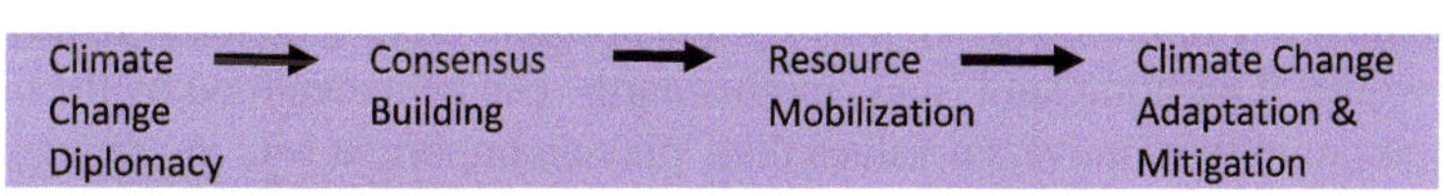

Fig. 5.3 Liner flowchart shows consequences of climate change diplomacy

With the goal of bolstering national efforts via tie-ups with international affiliations in reducing and adapting to climate change, climate change diplomacy seeks to push foreign governments and/or multilateral organizations toward particular policies. The issue of climate change is quickly becoming a focal point of international diplomacy at the highest levels. This is a problem for Bangladesh since the nation is developing and so more susceptible to the negative effects of climate change. It's a chance for the country to take the diplomatic lead on a global scale as its citizens learn more about the problem and begin to take action. A number of government agencies will need to strengthen their capabilities in the area of climate change diplomacy if the nation is to make the most of such chances.

As the country's primary climate change agency, the Ministry of Environment and Forests (MoEF) has been leading the country's efforts at UNFCCC meetings and the annual conference of parties (CoP), most recently at Cop17 in Durban, South Africa. Bangladesh has taken a leadership position at the cops within the LDC group to which it belongs, and its minister and ministry staff, as well as expert advisors, have amassed great knowledge and competence in the different negotiating tacks over the years.

The Economic Relations Division (ERD) of Bangladesh's Ministry of Finance will need to learn more about climate change finance, as it is distinct from the Official Development Assistance (ODA) with which it is more familiar. This is because international finance for climate change is only just beginning to flow from the global to the national level. The connection between Bangladeshi officials and their counterparts in the same developed nations has to be significantly different when discussing ODA (where Bangladesh has to accept what is provided on the conditions on which it is offered) and climate funding (where Bangladesh can dictate some of the terms). To provide one concrete example, the least-developed nations argue that only grants, and not loans, should be used for climate funding.

The Ministry of Foreign Affairs (MoFA) has to deliver monthly briefings on climate change diplomacy to its embassies overseas and in the longer run should send certain younger officers for further studies in climate change diplomacy. To ensure effective representation of Bangladesh at crucial high-level international events, Bangladesh may consider designating a personal "special climate change envoy/adviser" in light of the future significance of this issue. Many nations, both rich and poor, have sent "special climate change envoys/advisors" to address the issue. Instead of scientific knowledge, diplomatic skills, particularly within the UN system, are required for this role. That said, a senior diplomat who has served in New York or Geneva would be an excellent choice.

The government of Bangladesh would do well to use the language of harmony in international discussions on climate policy. By using metaphors with upbeat meanings, it aims to emphasize group identity and promote unity with those in the developing world. "Our homeland" and "mother earth" can be understood by the international community as universal values that "play an important role in argumentation because they allow us to present specific values, those on which specific groups agree, as more determined aspects of these universal values." The international community must act in accordance with the principles and provisions of the UNFCCC and the Kyoto Protocol.

To make America a better place for all of its citizens, we need to recognize that by helping others, we are also helping ourselves, confront climate change head-on, and cooperate to find solutions. There is evidence that Bangladesh's involvement in multilateral talks is a sign of the country's growing status in the international community. To deal with the challenges presented by climate change, Bangladesh desperately needs the West's financial and technical help. Bangladesh prefers to talk about the collective need of developing countries rather than its own specific interests while making political arguments.

Bangladesh has to fortify its bilateral ties with its neighbors so it is not left alone in the discussions, and it should also join a regional alliance to lend its voice to the need for more funding for preparations for the effects of climate change. However, it is important for Bangladesh to keep up its reputation as a trustworthy and helpful member of the international community. Its actions may be restrained by a desire to avoid public humiliation. Diplomatic efforts concerning climate change may be led by a policy cell established inside MoFA. Training diplomats stationed in MOFA's embassies across the globe can help bring attention to Bangladesh's environmental issues at home and abroad.

Not more charity or help is what we're after; rather, we want the world to take notice of Bangladesh's environmental significance, which is why we're focused on it. There should be a strong emphasis on technical expertise and knowledge among all delegates to international negotiations and conferences. Members of a delegation should be selected with care, taking into account their knowledge, experience, and institutional memory. Bangladesh's foreign policy since gaining independence has relied heavily on multilateral diplomacy, and this trend shows no signs of abating. Bangladesh has been an active participant in the multilateral system's pursuit of global political and socioeconomic stability and security. To this end, Bangladesh will actively participate in international fora, particularly the United Nations system and its specialized agencies, in order to advance security, international law, and development.

Due to its universal membership and its guiding charter, the United Nations plays a pivotal and important function within a multilateral system of government. Bangladesh looks to the United Nations to promote the development agenda relating to underdevelopment and the elimination of poverty because it acknowledges the need to address the urgent social and economic concerns of the developing countries. As a member of the multilateral forums, Bangladesh has a responsibility to advocate for the peaceful, legal, and UN-mandated settlement of international conflicts. It is important to Bangladesh that the United Nations and multilateralism be strengthened; hence, the country is backing any and all efforts to do so. Bangladesh's leadership positions in the NAM and OIC require it to play a vital role in pushing forward the southern development agenda. Considering the new dimensions of diplomacy that may thrive in the forms of bilateral diplomacy, track-II diplomacy, multilateral diplomacy, and cultural diplomacy, the author's own diagram in Fig. 5.4 might be viewed as an alternative to promote climate change diplomacy.

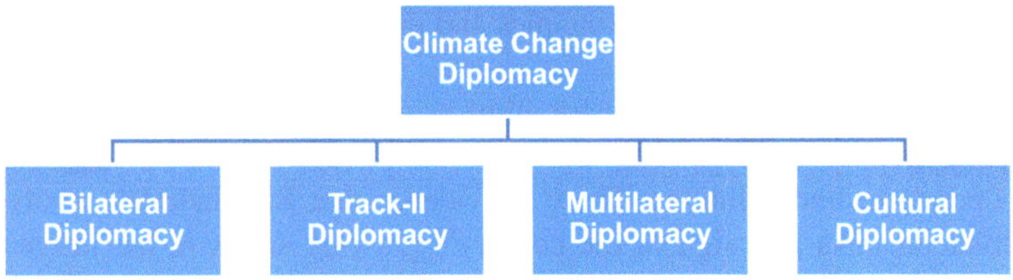

Fig. 5.4 Different means of climate change diplomacy

Toward the top, a plan is laid forth for Bangladesh that may be used to applaud diplomatic efforts that result in a win–win deal. To lay out the plan, including four stages, the goals, methods, and tactics that will lead a nation like Bangladesh to the greatest potential end will be shown. Aims and objectives, study and education for MoFA officials on the current global emphasis, competition, and the other party, strategies, and tactics are all defined. The Ministry of Foreign Affairs (MoFA) should also prioritize the following steps for result-oriented cc diplomacy:

- Don't worry about the individuals, worry about the issue. Relationship maintenance between governments, or between governments and international organizations/alliances, is the focus here.
- Investigate the thinking behind every negotiating stance (also, look for hidden agenda and personal motivators).
- Consider potential compromises and other paths forward (track-ii diplomacy or cultural diplomacy for unfolding the common interest of two nations).
- Consider only concrete examples (proactive and strong vocal in the international negotiation platforms gaining confidence and support of other climate change victim nations).
- You should be aware of the most promising substitutes for your first negotiating goals. (highlight the most dire consequences of climate change to a worldwide audience, while also highlighting Bangladesh's ability to adapt to these changes).

Those nations most at risk from climate change, such as Bangladesh, must take decisive action if we are to avoid a stalemate. Because climate change is becoming a national security concern, our governments must react to it like they would to any other national security danger. One way in which they might do this is by engaging in what could be dubbed "anti-diplomacy" to emphasize the need for a real global deal soon. Sometimes people from smaller or poorer countries may have a greater impact on the global stage if they are seen as competent and renowned experts in their field. If Prime Minister and administration are ready to take the initiative on the problem of climate change diplomacy, Bangladesh has a good chance of producing a globally renowned expert on the subject within the next few years.

By leveraging its many resources, including its capable officials, ministers, and a committed prime minister, as well as its many experts and NGOs that are involved both domestically and via extensive international networks, Bangladesh has the potential capability to play a leading role among the fellow-vulnerable countries on the issue of climate change. Bangladesh should develop a plan to play a proactive role at the international level in the next years to alleviate the suffering of victims of climate change caused disasters drawing the global help in the areas of mitigation and adaptation, given the importance of climate change diplomacy.

5.7 Community-Based Disaster Management and Local Government?

According to the United Nations' International Strategy for Disaster Reduction (UN/ISDR), a "disaster" is "any event that causes a substantial disruption of the functioning of society and poses a significant, widespread threat to human life, health, property or the environment, whether caused by accident, nature or human activity and whether developing suddenly or as a result of complex, long-term processes." Bangladesh is, without a doubt, the most disaster-prone country in the world. Repeated tragedies destroy the economic foundations of the impoverished and diminish their economic potential.

Humanitarian coordination that is both efficient and effective prioritizes meeting the most pressing needs of those who have been displaced by catastrophes in the most effective way possible, given the available resources. By preparing for and incorporating disaster management actions (risk reduction, response, and recovery) into local and national development plans, we can lessen the effects of disasters and the vulnerabilities of our communities. Coordination is a clear need for disaster management. Because of its efforts to protect its citizens from natural disasters, Bangladesh has received praise and recognition across the world.

Over the course of the past several decades, the government of Bangladesh has made a number of important initiatives to establish national and union-level institutional frameworks for efficient and organized disaster management. Developments in legislation, in addition to strengthening institutions, help reduce the suffering of catastrophe victims in Bangladesh. The government of Bangladesh has formulated a set of apparatuses, beginning at the national level and working their way down to the local level, to maintain proper coordination among the relevant ministries, departments, line agencies, Local Government Body, and community people and to ensure their proper functioning in order to alleviate the plight of the people. The Standing Orders on Disaster (SOD) serve as a blueprint for making these safeguards as effective as possible. According to SOD, many disaster management committees have been established, from the Prime Minister-led National Disaster Management Council to the union disaster management committee (UDMC), which reports to Chairman of the Union Parishad.

In line with the SOD, the UDMC will have 36 members, while its chair may appoint up to three (3) additional members and establish working groups and task forces as needed to address specific issues and concerns.

The UDMC has been given the responsibility of managing disasters in rural areas, including prevention, protection, response, and recovery. The Union Disaster Management Council (UDMC) is responsible for conducting a union-level hazard, vulnerability, and risk analysis and developing a risk reduction action plan, as well as keeping local residents informed and empowering them to take practical measures to reduce risk at the household and community levels (RRAP). UDMC will promote coordination among the development agencies and service providers via quarterly coordination meeting and take decision on implementation of the action plan for risk reduction as well as assess the status of the risk reduction action plan. Its goal is to help communities get the money they need to put their risk reduction strategy into action.

Unfortunately, locals, especially those in marginalized communities, have extremely little access to UDMC debates and decisions. As a result, it's safe to assume that the local vulnerable group's members are woefully uninformed about the disaster management committee's position, duties, and functioning. No evidence was found suggesting an involvement for UDMC in the events leading up to the accident. Community consultations showed that disaster risk management is not yet fully incorporated into the different Union Parishad activities, and this is generally accepted as the case. Disaster management committee leaders and managers are not subject matter specialists, but they have been given the authority to organize and administer disaster relief activities by the SOD. When political leadership at the local level is absent from disaster management, the interests of the people and their feeling of responsibility are not represented.

The citizens of Bangladesh who are most at risk from natural disasters are often denied information about the Upazila and district administration's disaster management programs, including details on the budgets allocated for each. However, the Union Chairman's position as the chairman of the union's disaster management committee has only existed on paper, and the committee's other members lack adequate training for their duties. In addition, Upazila Nirbahi Officer and Upazila Project Implementation Officer, who are in charge of the government's disaster management resources, are not enthusiastic about helping the union-level disaster management committee do its job.

Regular UDMC meetings before, during, and after a disaster; the creation of a designated fund for disaster risk reduction; the installation of disaster warning stations in each UP office; the building and maintenance of a disaster shelter center within the UP complex; the organization of a volunteer team under each UP for emergency response; the introduction of training on disaster preparedness and emergency response; these are all necessary for the UDMC to effectively address the challenges of disaster preparedness. In addition to these, the success of UDMC is a result of community participation in assessing hazard vulnerability and resources, developing plans, and putting those plans into action via mitigation and preparation measures.

Community members who take part in UDMC events are more likely to feel empowered to take on their share of the local development and disaster preparation tasks that are theirs. With the help of UDMC's capacity development and public awareness initiatives, communities are able to raise their level of engagement and, ultimately, continue their own preparation and mitigation efforts. The inclusion of two women in each UDMC is inadequate to guarantee that women's needs and capabilities are reflected, despite the best efforts of concerned government ministries like the Department of Disaster Management, NGOs/INGOs, and the UDMCs themselves. When it comes to women's ability to join and have an impact on UDMCs, neither evidence nor analysis is available. There is a lack of documentation about the aid and capacity building provided to UDMCs.

It is hard to acquire this data or to determine where disaster management committees are or are not working, despite the fact that several efforts are underway to increase their effectiveness. Union and Upazila elected officials play crucial roles in disaster risk reduction operations at the field level, while national resources are dispersed via district administration. However, the amount of granted resources is inadequate to fund the projects in this highly populated nation where communities are affected by many dangers all year long.

Good quality information on "disaster occurrences" must be established in collaboration with the government in order to rapidly determine their scale and any deficiencies in the government's ability to react. There has to be significant lobbying for the significance of releasing information quickly (regardless of the need for help) and for the establishment of a platform in which information created by non-government actors may be disseminated widely. It is important to take into account the legality of commencing coordination by planning out how to initiate a coordinated response to an occurrence that empowers Union Parishad, the first responder in a crisis [22]. The international humanitarian community must maintain its ongoing efforts by increasing the number of community-focused activities that directly connect to coordinated response readiness. The United Nations, INGO's, and the donor community have all been working to improve the situation recently. It is important that these initiatives work together.

5.8 Disaster-Induced Migration: Need Appropriate Strategies to Encounter

Cyclone Sidr hit the nation in 2007, killing 3447 people and forcing 650,000 to flee their homes (official record). Although Cyclone Sidr had a minimal effect on Bangladesh's GDP as a whole (the storm was responsible for a loss equal to 2.8% of the country's GDP), its impacts were felt most keenly in a small number of districts [23]. After Cyclone Sidr, there was an uptick in movement patterns, which threatened the safety of the human population. Good or bad governance in Bangladesh can be seen in the country's approach to planning for the future, raising public awareness, engaging in central government programs, working with non-governmental organizations as a development partner, effectively handling emergencies, and performing judicial and extrajudicial duties. Hence, disaster-affected migrants, as a socially excluded group, have practically no access to political benefits.

It is the duty of the GoB to provide aid in the form of social assistance and social services to help locals deal with a variety of problems such as those caused by natural disasters, extreme poverty, old age, widowhood, vulnerability, illness, unemployment, workplace injuries, and disabilities. But the government agencies that actually provide services have persistent issues to deal with. Because of this, it has not shown to be a reliable organization capable of providing enough social support and related services. Furthermore, official social security activities are insufficient to meet preexisting societal contingencies, leading to inadequate protection for disaster victims and migrants forced to flee their homes because of such catastrophes.

Unemployment, health risks, population increase, and inflated prices for basic necessities are creating hardships for the general public, especially catastrophe victims and migrants. Prospective migrants' future hopes and fears often center on their capacity to make a living in a certain climate and environment. The effect of increasing environmental pressures on migration is a question with no clear answer. Environmental degradation has a disproportionate impact on subsistence livelihoods in rural regions. Not only are women made more susceptible by the challenges posed by a rapidly changing environment, but their involvement in development efforts is also hampered by socioeconomic standards. The ever-expanding slums of Dhaka are a visible sign of the city's displaced population. A community's preparedness for a catastrophe may improve if its members have access to resources and knowledge. The capacity of the impoverished to cope with calamities is hindered by their lower levels of education and restricted access to information. Researchers have looked examined whether or not individuals living in disaster-prone regions have access to services despite their low income and precarious way of life.

Dhaka's slums and squatters are home to almost 3 million people who often go without basic services. Dhaka's low-income population live in slums and squatter colonies. Slums and squatter colonies are rife with conflict (quarrel, collision, brawl). This irritates city citizens, particularly adjacent residents, office employees, and kids, due to the noise and violence it generates. In addition, many people who live in slums engage in immoral activities such as prostitution, drug trafficking, hijacking, mugging, and so on. There is a threat to the city's cultural and social settings due to these actions. Also, physical violence is one kind of social conflict, which encompasses a wide range of other undesirable social interactions that may arise within social relationships.

Cyclone Sidr in 2007 highlighted the inadequacies and ineffectiveness of Bangladesh's disaster response. It took the government of Bangladesh more than a year to rebuild the damaged embankments in the worst-hit areas. Lots of individuals had to leave their houses because of the disaster. No sufficient rehabilitation efforts were made for the uprooted population. On top of that, response and rehabilitation programs aren't being held to account or monitored effectively. Recent natural catastrophes in Bangladesh have been met with widespread reports of ineptitude and corruption on the part of local disaster management officials in charge of relief and reconstruction efforts.

Problems related to massive population moves inside the nation, such as strain on existing limited resources or living in slums in ecologically dangerous places, may be a humanitarian concern, and estimates of the number of such movements are difficult to come by. Any future discussion of migration concerns must include consideration of environmental migration and those who have been forcibly forced from their homes due to environmental factors. Furthermore, it is suggested that these ideas be included into discussions about environmental concerns. The Bangladesh Bureau of Statistics, together with other research institutions and educational institutions, might play a pivotal role in establishing a solid scientific foundation to guarantee that the trend of disaster-induced migration is effectively followed. It's crucial to bring light to the environmental, social, economic, and political aspects of the problem among the public and political elite. The government and the media should go forward to raise awareness among the appropriate parties so that they can develop a unified strategy to tackle the issues. Civil society organizations are crucial in raising people's awareness about important issues [24].

Migrants forced to flee their homes due to natural disasters require legal protections, either as a stand-alone treaty or as provisions inside existing international agreements. The government of Bangladesh has to take the necessary steps to legislate the problem so that the displaced population may experience safety and stability. The city's ecology would benefit from the government cracking down on lawbreakers in order to make life there more pleasant for everyone. When contemplating the uprooting of whole communities, it is imperative that the appropriate bodies within the United Nations system and other major humanitarian assistance organizations be given the authority to help those who have been uprooted from their homes due to environmental factors. To do this, an international procedure recognizing these people is necessary. For the Bangladeshi government to successfully engage with its foreign counterparts, it must grasp the need of using forceful diplomacy.

Finally, ideas need to be developed and institutions need to be improved or established in order to aid the influx of people who are fleeing their homes because of environmental threats on a global and national scale. Nationally, this may entail bolstering and encouraging cross-departmental collaboration between relevant agencies (e.g., the Department of Disaster Management, the Ministry of Home Affairs, the Ministry of Environment and Forest, the Ministry of Social Welfare, the Ministry of Disaster Management and Relief, and so on) to address the issue holistically.

The social and economic losses individuals suffer need to be better understood so that humanitarian interventions, including community relocation, may be properly structured. Finally, the many types of environmental migrants should be taken into account in any new policy. Migration should be conducted in a way that is safe, legal, and orderly, and this requires cooperation between institutions in both the country of origin and the country of destination. Following a catastrophe, a strong restoration strategy is necessary to keep people from leaving the region. In the immediate aftermath of a catastrophe, it is crucial to begin a rehabilitation program that will provide survivors the tools they need to rebuild their homes, their communities, and their livelihoods. Most importantly, the law and order situation in the slum area must be maintained effectively in order to avoid social strife, and the government of Bangladesh's social safety net scheme must be extended to include catastrophe victims who have relocated to urban regions.

The potential scale and extent of environmentally induced migration make the topic of migration caused by disasters an urgent one from a policy perspective. Stresses from a changing climate, together with other factors, such slow and sudden changes in ecosystems, are already a major factor in migration. The government of Bangladesh must begin making preparations immediately to relocate and resettle the affected people. It is crucial to understand how changing environmental circumstances impact individual and collective choices to migrate as the need to discover viable adaptation strategies for climate change becomes more pressing.

To lessen the chances of conflict or even possible conflict centered on migration, a fresh level of policy and scientific focus is needed to find the policy solutions to smooth the road ahead and prevent conflicts over natural and social resources. Further, for the sake of coexistence and to forestall social instability, the government must assume the duty to provide social security for the migrating people living in the slums. Bangladesh as a country needs to learn how environmental processes and environmental quality influence people's standard of life at home and influence their decision to move to the capital city or to metropolitan regions. Multilateral discussion may be required at the regional level to address, coordinate, and alleviate environmental constraints and migration resulting in part from climate change-related catastrophes in Bangladesh. Bangladesh's government should make the most of international fora in order to attract greater funding for disaster prevention and response.

5.9 Donor-Driven Disaster Risk Reduction Programme in Bangladesh: Question of Dependency

When a tragedy strikes in Bangladesh, donors are quick to respond. As the primary venue for government and donor agencies to make strategic decisions and exchange ideas and information on disaster management, the Local Consultative Group on Disaster and Emergency Response (LCG-DER) includes participation from all important stakeholders. Within the larger context of disaster management (risk reduction, preparation, relief/response, and recovery/rehabilitation), the DER has the responsibility of coordinating the activities of national and international parties. Secretary of the Ministry of Disaster Management and Relief serves as the LCG-chair, DER's with the UN Resident Coordinator serving as the cochair [25].

The LCG-DER authorized a new humanitarian coordination framework in January 2012, after an extensive examination of the previous system in Bangladesh. The Humanitarian Coordination Task Team (HCTT), comprised of eight humanitarian clusters [26] (Food Security, Nutrition, Health, WASH, Education, Shelter, Early Recovery, and Logistics), has been tasked with ensuring a coordinated, timely, and inclusive response among partners in their respective sectors, in close collaboration with the relevant line ministries, in support of the government's response plan (Fig. 5.5; Table 5.5).

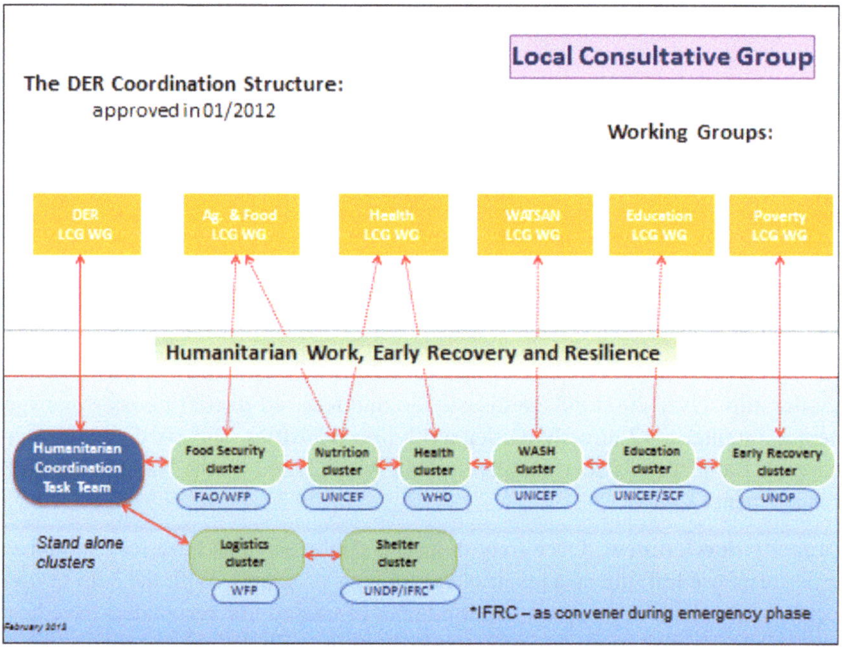

Fig. 5.5 DER coordination structure. *Source* LCG-DER Coordination Tree, ERD & MoFDM, Dhaka (2012)

Table 5.5 Humanitarian clusters and relevant government ministry

Cluster	Cluster Lead Agency	Government linkage
Food security	WFP and FAO	Ministry of Food
Nutrition	UNICEF	Ministry of Health
Health	WHO	Ministry of Health
WASH	UNICEF	Department of Public Health and Engineering
Education	UNICEF and Save the Children	Ministry of Education
Shelter	UNDP and IFRC	Ministry of Disaster Management and Relief
Early recovery	UNDP	Ministry of Disaster Management and Relief
Logistics	WFP	–

Source LCG-DER Membership, ERD & MoFDM, Dhaka (2012)

Donors reportedly assist Bangladesh in building up its infrastructure for disaster management and risk reduction by providing funding for infrastructure improvements and training for disaster management personnel. These expenditures aid in creating a more favorable catastrophe situation. The pillars of this shift are donors reportedly assist Bangladesh in building up its infrastructure for disaster management and risk reduction by providing funding for infrastructure improvements and training for disaster management personnel. These expenditures aid in creating a more favorable catastrophe situation. The pillars of this shift are

A fresh approach: The Comprehensive Disaster Management Programme and the National Risk Reduction Agenda are the results of decades of experience in disaster management.

Leadership: Over 2000 village disaster committees, 40 district disaster management committees, 12 directly participating line ministries, and six donor partners make up Bangladesh's coordinating system, which is led by the National Disaster Management Council.

Capacity development: Over sixty thousand public servants have received emergency response training as a result of (a) training programs. With the help of civil society and volunteer groups, a larger number of trained first responders have been assembled; (b) institutions and policies: Institutions with defined missions serve as focal points for development partners. National institutions have been more proactive with the aid of donors, implementing early warning systems and mitigation methods; (c) civil society engagement: donor support and advocacy have helped empower civil society groups to strive for change [27].

A multi-hazard focus: More factors, such as the effects of climate change, are being taken into account in disaster management and risk reduction efforts. Bangladesh's performance may be gauged by the fact that the number of people killed by natural catastrophes has dropped dramatically. Hundreds of thousands of people have been killed by isolated incidents throughout history. Recently, however, the trend has been slowly decreasing, with the exception of two very powerful cyclones in 1995. Certainly, there are still problems to be solved. However, there are also challenges in sustaining attention during "normal" times between crises and ensuring that all regions of the nation have equivalent capability. It is an ongoing endeavor in Bangladesh, as it is everywhere, to mainstream catastrophe issues and include them into development planning.

Donors' post-disaster aid for emergency relief and reconstruction has fallen short of what developing nations like Bangladesh need to lessen their vulnerability to disasters and provide adequate resources to their governments and citizens so that they can pay for the clean-up and rebuilding that must occur.

In the developing world, natural catastrophes continue to be a leading source of death and destruction of property. Death tolls per event were several times higher in low- and medium-income nations than in high-income countries in a sample of big natural disasters from 2001 to 2012, and losses as a proportion of GNP were substantially negatively linked with per capita income. Most aid is delivered after a crisis has already occurred, despite mounting evidence that spending on prevention yields substantial returns. As a result of the moral hazard that comes with providing aid after a catastrophe, governments and people are less likely to engage in prevention when they know they will be provided for regardless of how much preparation they do.

Donor assistance alone will not be enough for developing nations. In most cases, the OECD Development Help Committee reports that humanitarian aid is only covering a tiny fraction (typically less than 10%) of catastrophe losses in recipient countries. On the other hand, budgeted funds may fall short of expectations. Despite initial pledges of billions of dollars, central reserve fund and foreign aid had barely reached 20% of that total two years after the 2001 earthquake in Gujarat, India [28]. In light of the fact that disaster aid as currently structured is unlikely to be sufficient to meet the needs of developing countries like Bangladesh in the aftermath of a catastrophe, the international community should reconsider shifting its focus from relief to risk management programs that combine public and private resources to maximize aid effectiveness and minimize losses. The potential of risk-transfer initiatives funded by donors is shown by three case studies.

Financial institutions, international agencies, non-governmental organizations, and donor governments are all becoming aware of the need to prioritize disaster loss prevention programs, but they lack the resources to contribute directly to the government's direct funding facility for a coordinated disaster risk reduction programme. Local non-governmental organizations (NGOs) work with international donors including SDC, DFID, UNDP, WFP, and UNICEF to carry out aid programs. Most DRR programs are developed and implemented without consulting local residents or government officials. Mismanagement of assets is seen. The aftermath of Cyclone Sidr prompted us to take attention. Donor communities responded independently to Cyclone Sidr because of a lack of political authority from the government. Donor agencies seldom engaged with local government and coordination levels were low. Having to rely on local non-governmental organizations (NGOs) might hurt their credibility since some of these organizations have been accused of recruiting charity recipients in exchange for financial support.

5.10 Governing Environment Enables Better Disaster Management

In political ecology and environmental policy, "environmental governance" refers to the process of identifying and establishing the structures and procedures essential for ensuring the long-term viability and adaptability of an ecosystem. It's vital to people's existence, especially for those who rely on the environment and natural resources to make a living, and it's intertwined with other facets of sustainable development. There is a strong correlation between natural resources and a wide range of social, economic, and political problems. To address this problem, environmental governance, which is “the coordinated administration of policies involving social and environmental actors,” brings together the expertise of many groups and organizations.

Given its location, hydro-geological features (such as the predominance of floodplains and low elevation from the sea), socioeconomic features (such as high population density, high levels of poverty, and overwhelming dependence on nature), and political and cultural institutions, Bangladesh is widely acknowledged as one of the most vulnerable countries in the world, highly vulnerable to climatic manifestations. Bangladesh is home to a wide range of ecosystems, and as a result, the country’s inhabitants use a wide range of land management practices, from the ancient to the cutting edge, each carefully tailored to the specifics of the land on which it is used. Insightful implications for vulnerability and depletion of the natural resource base may be drawn from this complexity of environment and consumption patterns. This country’s economic development and quality of life are hampered by a number of reasons, including its high population density, slow economic growth, lack of institutional infrastructure, heavy reliance on agriculture and agricultural goods, geographical locations, and others.

The government of Bangladesh took part in the general process of conserving the global environment in accordance with the Stockholm mandate 1972. In 1973, Bangladesh’s government passed the first Water Pollution Control Ordinance, and in 1977, it passed the Environment Pollution Control Ordinance, both of which were intended to implement the Stockholm Accord. The Department of Pollution Control Ordinance was founded in 1985 to implement environmental programs, and it has since been renamed and reorganized as the Department of Environment (DOE). It wasn’t until the passage of the Environmental Policy in 1992 that the concept of protecting the environment via national efforts was officially acknowledged and stated.

The development of environmental policy was influenced by a wide range of stakeholders and external causes. Each and every stakeholder, both internal and external, had an essential role in developing the environmental policy. Concerning environmental issues, in 1992 Bangladesh’s government adopted an Environment Policy. Environmental policy’s primary goals are to safeguard the nation from natural catastrophes, to preserve the environment, and to identify and regulate all activities that pollute and degrade the environment.

The People’s Republic of Bangladesh government has acknowledged climate change as a pressing problem, and efforts are being made to include possible response measures for mitigating climate change consequences into the country’s overall development planning process. Consensus is growing that Bangladesh, a nation that is particularly susceptible to the effects of climate change, would have to work even harder to achieve its development goals. In 2005, the People’s Republic of Bangladesh’s Ministry of Environment and Forest (MOEF) responded to a resolution made at COP7 of the United Nations Framework Convention on Climate Change by drafting a National Adaptation Programme of Action (NAPA) (UNFCCC). To achieve its sustainable development goals, the country’s fundamental approach to NAPA planning was to include stakeholders in negotiations over resource usage, allocation, and distribution, as well as to address environmental issues and natural resource management. Furthermore, in 2009, Bangladesh unveiled a Bangladesh Climate Change Strategy and Action Plan to address the risk of climate change by implementing adaptation measures using both domestic and international resources.

The Ministry of the Environment and the Forests is formally responsible for the whole environmental sector (MoEF). There are, however, a plethora of additional institutions that have a role in either the management or formation of the environment sector, either directly or indirectly. These include both governmental and non-governmental organizations. The Ministry of Environment is responsible for coordinating with other government agencies to ensure environmental issues are properly addressed in the nation's development strategy. The ministry is responsible for providing advice on government policy and coordinating the rollout of sector-wide action plans. All development efforts' effects on the environment have to be evaluated and tracked by MoEF. The Department of Environment (DoE), a specialized division of the Ministry of Environment and Forests, is responsible for carrying out environmental programs in the field. In addition, several government agencies in Bangladesh are tangled up in a foundation for tackling basic challenges of environmental management.

In Bangladesh's top-down decision-making system, the central government sets decisions and is followed by local governments and administration to carry them out. As a consequence, policies and processes do not accurately reflect the situation on the ground due to a lack of feedback from lower to higher levels in the decision-making process. On the other hand, there are not enough channels of communication between decision-makers, the public, relevant players (civil society members and non-state actors), and the media in the decision-making process. This makes it less likely that non-state actors will take action in response to regulations, and it also means that the community targeted by environmental protection programs is unlikely to have an active role in encouraging their adoption and use.

Existing environmental policy guidelines do not provide functional tools to address climate change; even environment policy does not make reference to the phrase "climate change" or its negative effects. The process of implementing environmental policy in Bangladesh does not include any kind of official or informal discussion between government agencies, notably the MoEF and DoE, and polluters. The relevant ministries lack the necessary institutional competence to carry out the different action steps.

Contrarily, neither the young MoEF nor the DoE has built the institutional competence to tackle environmental management and protection issues on a large scale. MoEF and DoE are unable to provide effective environmental governance due to a lack of openness and public input throughout the decision-making process. Lack of institutional capabilities, untrained human resources, lack of awareness, low community participation in resource management, a dearth of research, and a lack of coordination among various stakeholders are all hypothesized to contribute to Bangladesh's abysmal track record of environmental governance (governments, UN agencies, NGOs, private sector, and civil society).

The Ministry of the Environment and the Department of the Environment (MoEF and DoE) should implement a mitigation and adaptation strategy to create a centralized database and management information system to ensure the success of all national environmental policies (MIS). Climate change vulnerability analysis calls for a revamp of current environmental policies. India should speed up bilateral discussions and strengthen the Joint River Commission to resolve trans-boundary water concerns [29]. Involving and worrying those who can make a difference requires first launching a community awareness and information distribution campaign. Through these initiatives, researchers and policymakers may gauge public opinion on environmental issues, climate change, and adaptation. To improve the efficiency of political choices, institutional, administrative, and organizational changes would be required.

This would follow an analysis of the current organizations handling climate change concerns, such as national climate change committees, their level of representation, and the powers and duties that come with it. Services geared at establishing environmental governance would be provided with better sector-department coordination and integration as a goal. The local government's management of natural resources and the environment should be held accountable and efficient, thus efforts should be focused on figuring out how to make that happen.

Those who have a vested interest in the environment should play a significant role in streamlining environmental governance through activities such as data collection and sharing, policy deliberation and drafting, policy implementation, evaluation and oversight, and promotion of environmental fairness. Remember that NGOs and other civil society organizations are not only active participants in government, but also a driving factor behind increased international cooperation by actively mobilizing public support for international accords.

In environmental discussions in particular, civil society groups have long been seen as "partners" of the UN system because of the crucial role they play in service delivery and implementation. Most importantly, non-governmental organizations and the United Nations system should collaborate to support the government of Bangladesh's efforts to speed up the implementation of environmental legislation. This includes international finance and development agencies as well as all intergovernmental organizations and forums.

The creation of a national platform is a regular result of laws pertaining to disaster management. Platforms at the national level may facilitate communication between relevant government agencies and the many stakeholders involved in formulating policies to mitigate risks. Legislation and its execution at the national level may be bolstered by effective regional government. Support at the regional level is essential for the successful implementation of laws aimed at lowering risk. Only 55% of country replies to the UN-WCDR mentioned the strategic use of development policy to mainstream DRR, indicating that risk reduction policy remains on the margins. There aren't many nations that have integrated DRR policy into broader national development planning documents like Poverty Reduction Strategy Papers (PRSPs). Respondents who raised issues with the implementation of DRR rules cited the following factors as causes for weaker policy enforcement:

Concerned Ministries' Action Plan is off; Typical Mindset Regarding Relief and Recovery; Lack of Transparency; Absence of Penalty and Reward System; Lack of Pragmatic Ground-Level DRR Program; Lack of Awareness

In terms of DRR, there is no one entity that can handle all of the issues. To those who adopt a DRR mindset, catastrophes are complicated challenges that call for a group effort to solve. Even in the most basic emergency management scenarios, coordinating the efforts of the many different groups that may descend on a disaster zone to provide aid may be challenging. Relationships between organizations and across sectors (public, commercial, and non-profit, as well as communities) grow increasingly wide and complicated over the whole range of DRR. Disaster preparedness and response call for solid horizontal and vertical ties (central-local relations become important). The following is a summary of the respondents' opinions and recommendations for what further should be done to spread awareness about DRR.

Increase expertise within the relevant ministries on DRR; incorporate sustainability into DRR programming; promote an integrated approach involving relevant ministries; guarantee a strategic, "joined-up" approach to disaster relief. Action plan of concerned ministries; proper orientation about implementation; ensure capital punishment; mindset change; create awareness on the ground.

Institutional arrangements and procedures that allow agencies to participate in executing a unified plan are required to ensure that government and non-government agencies contribute to the solution of a common problem during disaster management's various phases (disaster emergency, recovery, and risk reduction). It is crucial to compile and collate all guidelines developed by the Bangladesh government that define the entitlements of affected people into a single document, which is readily available to all personnel in the field involved in emergency response, in order to provide humanitarian assistance and DRR programs effectively and equitably. Also, field staff must be given explicit instructions on how to use their judgment when dealing with situations for which rules have not yet been completed.

Coordinating and making "disaster risk reduction and response actions of all governmental and non-governmental entities object oriented and robust" are DDM's main responsibilities. Reality shows that humanitarian response organizations seldom have a common understanding of the issue at hand. Individually, they devise responses, and their aims and techniques are so dissimilar that they almost never converge on a shared goal. Another issue is that financing arrangements sometimes provide limited room for maneuver for different organizations. Therefore, it becomes very difficult to coordinate humanitarian interventions. In times of extreme danger, such floods and cyclones, it is essential to relocate the population. Despite this, there were not enough flood or cyclone shelters in these high-risk locations, and the ones that were there were much too tiny to house everyone who needed them.

Interventions in DRR, response, and recovery all required rescue operations, needs assessments, and the distribution of relief supplies to be carried out successfully. These folks need to be technically savvy and also considerate of women's, children's, and people with disabilities' needs. It was notoriously difficult to round together enough qualified workers to respond to an emergency. Local disaster management committees utilize the D-form to gather primary data on damage and needs. When it comes to collecting and documenting information, however, local disaster management committees are woefully inadequate. Data storage, compilation, and analysis are also not up to par, making it difficult to prepare for a quick reaction.

Based on what we've learned, we know that promoting community-based disaster risk reduction and integrating risk reduction elements into the emergency response are critical steps toward enhancing communities' capacities to protect life and assets and improve their ability to mobilize resources during emergencies. It would be ideal to have local disaster management organizations (DMOs) like UzDMC and UDMC carry out DRR and emergency response activities, and this would need bolstering the DMCs' capabilities via the provision of different inputs, including training.

Having a preexisting "volunteer squad" in the community may help to rapidly organize human resources for evacuation, rescue, and relief distribution in the event of a catastrophe. Given the unpredictability of emergency situations, it is crucial to have a framework in place that enables DDM to reallocate money to ensure that aid is delivered as soon as feasible. Allocating appropriate resources in a more methodical manner to meet the expenditures of the operation is necessary to properly handle the demands during an emergency.

The legal and regulatory framework does not fully address recovery. Ministries in Dhaka and in the provinces often lack an understanding of the NDMA, NPDM, and SODs. This is especially important for the SODs, which define the roles of all federal, state, and local agencies in preventing, mitigating, and responding to disasters. A recent research revealed that many DMCs do not keep a copy of the SODs on-site.

The government operates a variety of DMICs, or disaster management information centres. To consolidate Bangladesh's expertise in disaster management, DMIC has recently developed the Knowledge Library. To make the government's disaster information management cell's repository the "go to" place for catastrophe-related data, it will need persistent work and attention from all parties involved.

Between 2007 and 2012, significant progress was made in the area of disaster management information products; regular status reports were generated in the wake of catastrophes; and the development partners contributed to the new emergency planning tool. However, the system is not equipped to examine the reaction, much alone provide accurate and up-to-date gap analysis, for recurrent catastrophes like floods.

Skillful management of information resources is required for coordination to effectively find and eliminate redundancies and gaps. The coordination of relief efforts relies heavily on information management, a field that is often overlooked and poorly understood. Disorganized data is a big barrier to education. There is little sign of an open-door policy or culture of openness in the field of disaster management, and there are few chances for the kind of critical lesson learning that may help advance efforts to strengthen disaster policy and practice.

Government and development partners give funding for catastrophe preparation, response, and recovery. A current gap analysis is required to determine whether a catastrophe is being sufficiently supported. Because of this gap, advocating to individuals with the power to release funding is ineffective. Amounts of donor cash received for relief have been included in DDM's Annual Report, although this data is based on information supplied by OCHA's Financial Tracking Service (FTS), and so does not reflect where development money is spent on preparation, recovery, or small-scale activities. Spending on each catastrophe by DDM or other government agencies could not be analyzed. Although it was mandated by law under the National Disaster Management Act, no such Disaster Management Fund has yet been formed. If money came from it, coordination might happen more easily.

Due to the prevalence of "micro-level" catastrophes that go unreported at central level but have a profound impact at the resilience of a community, the Local Disaster Management Fund is crucial. While the DRR fund may serve as a learning tool in this respect, it does not provide a complete blueprint for setting up a local disaster management fund. The Deputy Commissioner and the Upazila Nirbahi Officer are two of the most hesitant local officials when it comes to creating and using this emergency relief fund. It's not obvious in practice. Changes would have to be made, notably in the areas of decentralizing decision-making and including real community accountability systems.

When it comes to disaster planning, response, and recovery in Bangladesh, women continue to be underrepresented. In one example of a policy shift attributable to gender analysis, it was discovered in the years following 1991 that women were less likely to relocate to cyclone shelters than men were because of a lack of privacy, prompting authorities to ensure that all future shelters feature separate areas for men and women. This policy decision's conclusion was not disclosed. There isn't a major lack of gender parity in the existing set of laws and policies. Women are recognized as a marginalized population, yet there is a lack of awareness about the disparity between male and female survivors' abilities and the effects of catastrophes.

The literature study showed that development partners only consider gender in two aspects of their programming: women's engagement and women being the primary beneficiaries. There was a severe lack of literature that conducted a vulnerability assessment of how men and women, girls and boys, all contribute differently to disaster management and have unique requirements. We need to go beyond our very simplified gender analysis. Oversimplifying and leading to a multitude of unintended effects that hurt women, rather than help them, is the belief that men would abuse cash or in-kind offerings; hence, it is preferable to donate to males. There is a lack of attention to gender equity in DRR proposal making. Neither a gender analysis nor any mention of gender-based violence is included in the current coordinated assessment instruments in Bangladesh.

Gender analysis has been lacking in disaster management policy and practice generally. Disaster management efforts have not taken into account women and girls' unique needs and strengths. Since a gender lens has not been used to monitoring and evaluation indicators, knowledge on how disasters affect women and girls, and how they could contribute to prevention, response, and recovery is limited. The Bangladesh Meteorological Department (BMD) has benefited from the substantial scientific progress made in cyclone forecasting over the last decade. In a similar vein, the FFWC has increased the period of advance notice for river floods from 3 to 5 days.

Volunteers with the Cyclone Preparedness Programme (CPP) are said to be really excited to help out their neighborhood; they are in constant contact with Union Parishad leaders, and the majority of them are already well-known to the locals. CPP has shown its flexibility by growing in the southwest after Cyclone Aila hit there. This is an area seldom hit by cyclones. The early warning system has been upgraded to add IVR capabilities. The lack of an early warning system for natural disasters including flash floods, landslides, and drought is especially pressing in metropolitan settings. The CDMP is now conducting trial operations in this field. Riverbank erosion modeling efforts should be used to flood mitigation efforts.

Moving to cyclone shelters, rather than just hearing the warning, is the most important factor in determining survival during a cyclone. Protection considerations and the specific needs of women and girls and those living with disabilities were included in the 2009 Report on Cyclone Shelter Information for Management of Tsunami and Cyclone Preparedness Planning as a result of lessons learned from the post-Sidr cyclone shelter construction boom, which saw a number of stakeholders rush the building of cyclone shelters against tight donor funding cycles. Local governments and non-profits may now access information on the location, characteristics, and condition of all storm shelters in their areas thanks to the database.

The number of cyclone shelters available to the "at-risk" population is insufficient, and in certain situations, the size of these shelters is insufficient as well. Even though cyclone shelters are easily available, it is important to recognize that various families may choose alternative sheltering choices, such as seeking refuge with neighbors who have more sturdy homes. Only 12% of persons in hurricane Sidr sought refuge in designated shelters. Cyclone shelters must be easily accessible in the event of an actual storm [30]. In the past, cyclone shelter policy has only concerned itself with the building and upkeep of shelters, but this has lately been broadened to include the creation of accessible routes to the shelters, so that people of all ages and abilities can get to them in the event of a cyclone.

There has to be more than just a vertical or horizontal chain of responsibility. People's voices and capabilities in disaster planning, response, and recovery have been inadequate, and the gap between policy and practice continues to hamper disaster management. Despite its reputation as a world leader in disaster management, Bangladesh's monitoring and assessment methods are inadequate, leading to a lack of data.

There is a dearth of inspiration for a culture of learning and development in disaster management due to a lack of motivation toward continuous improvement and critical reflection on actions. The policy level recognizes new and growing threats, but there has been little implementation. In actuality, disaster management initiatives continue to center on floods and cyclones, and the disaster management community lacks a sufficient grasp of how they should be planning for and reacting to Bangladesh's complex and evolving hazard environment. Given the country's large urban population, reeducating people on the dangers of natural disasters is perhaps the biggest issue facing Bangladesh today.

Bangladesh has seen a dramatic increase in the number of city dwellers, with 32% now compared to only 7% in 1971. It wasn't until 2012, however, that the possibility of a catastrophic urban event was seriously considered. Compared to the progress made in the face of natural disasters in coastal regions that made in the face of urban catastrophes seems quite discouraging. Its reach is also somewhat restricted (in some 50 schools in Dhaka and Rangpur city and some slums in Dhaka). Around 43 million people already live in cities, and this number is expected to grow. Many of them, like those in rural areas, are at risk from natural catastrophes and need our help [31]. Numerous Sustainable Development Goals (SDGs) include measures to lessen disaster risks (e.g., Goal 11, 13, 14, etc.). Creating a city that can withstand natural disasters is one of the SDGs. Long-term planning to lessen the load on cities by, for example, decentralizing the provision of essential services and decreasing the frequency with which natural catastrophes strike is essential. Some of the gaps in this area include the absence of land zoning and control of private real estate development, a contingency plan to best use the resources, risk assessment, collection and distribution of data, and so on, as stated in the Hyogo Framework progress report of Bangladesh.

Authorities have given their stamp of approval to the Early Recovery Guiding Principles. No concrete takeaways related to recovery policies were identified. These SODs talk of becoming well. Government officials and aid organizations have both committed to starting the rebuilding process quickly. Despite the fact that DDM's primary mission continues to be humanitarian aid and recovery, this is the monetary and non-monetary aid given to victims of a catastrophe. There have been some positive developments in DDM's efforts to carry out rehabilitation initiatives during the last four years. After the devastation caused by Cyclone Mahasen, the government and development partners pooled together about 490 million BDT to fund house reconstruction and other early recovery efforts [32]. The government used the Climate Trust Fund to construct cyclone-proof housing in Aila's hit districts. Although early recovery is beneficial to disaster risk reduction, the notion has not yet been included into Bangladesh's Comprehensive Disaster Management Model.

The first and main key hurdle to disaster risk reduction action is the desire and political commitment of numerous entities, including the government and large funders. The proportion of the national budget allocated to DRR should be carefully analyzed. Furthermore, we must examine the number of donors supporting DRR initiatives and the proportion of their budgets that goes toward DRR interventions. It's been noted that nations with the lowest incomes are also the most at risk from catastrophic weather events. There is considerably less room for the country as a whole to recover. No matter how much the government may want to help, it just doesn't have the means to do so.

The government often debates whether to spend on disaster-related problems or meet the fundamental necessities of the people. There has to be a concerted effort from the international community to assist this country secure its citizens against harm. I believe this issue might potentially be resolved by raising awareness and teaching policymakers that investing in DRR activities can save otherwise-lost development endeavors. It is important to strengthen the connection between progress and DRR. Lack of coordination between different DRR entities is also a key problem, and this is true across countries of varying economic position. DRR field activities in

Bangladesh were the topic of a recent research. A greater number of DRR organizations were located in certain regions than in others, even though the latter were found to be more at risk and deserving of more resources. A comprehensive study revealed that these locations were the most accessible and noticeable. Guests of honor, such as donors, like to congregate in these sections. Misallocation of resources prevents investments in areas that may have the greatest impact.

Due to decreasing funding from donors and other organizations, this issue requires significant consideration. Coordination between relevant authorities, coordinated by the government and aided by OCHA, is one approach to addressing this challenge. Together, you and your partners must do a thorough vulnerability analysis and then decide where to intervene and how to help those most in need. The government has been seen to be taking some initiative in this area. To identify susceptible locations and necessary treatments, the CDMP (the biggest DRR initiative in Bangladesh to date) is conducting a mapping project. This will be made publically available on the official government website so that anybody may use it as a resource for organizing interventions.

Another important difficulty I've seen is the absence of standardization and uniformity in DRR treatments. Organizations are using a wide variety of training modules and reference materials to improve the community's DRR capabilities, as was shown in the field. One day of training was offered by some companies, whereas the same subject was covered in depth over the course of many days by others. Therefore, the community's differential capacity improves. After becoming aware of the problem, the government of Bangladesh is now working on creating a consistent training program for the many parties involved.

Numerous Sustainable Development Goals (SDGs) include measures to lessen disaster risks (e.g., Goal 11, 13, 14, etc.). Creating a city that can withstand natural disasters is one of the SDGs. Long-term planning to lessen the load on cities, both through decentralizing basic services and decreasing catastrophe occurrences, is essential for this [33]. Some of the gaps in this area include the absence of land zoning and control of private real estate development, a contingency plan to best use the resources, risk assessment, collection and distribution of data, and so on, as stated in the Hyogo Framework progress report of Bangladesh.

All sectors of the economy, including food production, health care, education, and transportation, are vulnerable to natural disasters. One of the primary industries hit by the rising frequency and severity of disasters is agriculture, which accounts for roughly 29% of the country's GDP and employs around 63% of the entire labor force (BBS, 2006). Damage to health and educational facilities from natural disasters like floods and cyclones disrupts the supply of essential medical services and makes it difficult for students to continue their education. The transportation network is likewise vulnerable. Reconstructing infrastructure after a natural catastrophe is a time-consuming process. People with less money and resources are more likely to suffer serious consequences in the event of a catastrophe. People who are impacted will likely lose all they own. This means that extreme poverty levels tend to rise in the wake of natural catastrophes, include catastrophe risk reduction initiatives in the development agenda to help vulnerable populations, and make the development process more fair (for the poor and the affluent alike).

Bangladesh is particularly susceptible because of its extreme poverty. High exposure and vulnerability combine to put Bangladesh at danger, since over half of the population lives on less than USD 1.25 per day. With rising sea levels and more intense storms and severe weather events, coastal ecosystems and populations in Bangladesh will be hit the most by climate change [34]. The people of Bangladesh are particularly vulnerable to natural disasters since so many of them live in low-lying coastal regions. High wind and tide velocities along the coast have increased the potential of harm to coastal ecosystems including coral reefs and mangroves.

Located at the mouth of the world's biggest delta, Bangladesh's Southwest Coastal Region is blessed with a rare ecosystem, rich soil, and other natural resources. Rivers, estuaries, the biggest mangrove forest in the world (Sundarbans), and a brackish water regime make up this tidal wetland. There is barely one meter of elevation here. The sandy deposits pushed down by the Ganges' many tributaries help build this delta of loose soil. Even though these rivers are no longer linked to the rivers farther upstream, they continue to exist as transporters of overflowing rain water. In addition, they are linked to one another [35]. Their currents are controlled by the seasonal monsoon rains and the tidal flow from the Bay of Bengal. Large amounts of sediments, including decayed forest debris, float on their backs. In the water, the decomposition of the 3.5 million tons of forest debris that fell from the Sundarbans creates vast amounts of organic food that fish and other aquatic organisms may use to thrive [36]. Because of the unique characteristics of the soil, this area is one of the most fertile in the world. But in the southwest, crop failure was prevalent because of the inundation of salt water and the monsoons.

A better existence is a fundamental human right. However, this group is being discriminated against. Affected people and experts agree that donors constructing embankments in the wrong design without understanding the region's unique geophysical characteristics, India constructing the Farakka embankment while ignoring the downstream Bangladesh, illegal or poorly designed constructions over rivers and canals, river dying, and so on are the main causes of these calamities. The corruption of project implementers, political leadership, pursuit of narrow political benefit, and shrimp farming are all factors that have made the situation worse.

What we discovered is that politicians on both the national and municipal levels do not want to find a solution. They constantly keep an eye on the ballot box. Even if one side of the argument realizes the issue has been solved, the other believes that the winner shouldn't be given the chance to become a fan favorite. In addition, between 60 and 70% of those impacted in Manirampur and Abhaynagar identify as Hindu. For some members of these minorities, the fact that they are religious makes it difficult for them to find a solution. Some individuals hope that by encouraging these Hindu families to go to India, they would be able to benefit politically. The public has no interest in being helped, receiving preferential treatment from anybody, sharing in corrupt profits, or holding positions of political power. To put it simply, all they want is to make it through this life intact. Their right to be evacuated from this man-made disaster is guaranteed under the constitution.

Historically, androcentric civilizations have maintained that gender imbalance makes women more susceptible to disaster-related loss and mortality. Within this created reality, women have less access to health care and food, less freedom of movement, and less opportunities to participate in political and economic decision-making and are subject to weaker human rights protections in Bangladesh's legal, commercial, and political institutions. When women are excluded from decision-making in public, they are less likely to have access to knowledge that might save their lives in times of crisis.

Even if women are aware of impending storms (through the media, community volunteers, government agencies, NGOs, or word of mouth from neighbors), the study found that they still relied on the (male) household head to make the final decisions, such as whether or not to seek refuge in cyclone shelters. Women's contributions to disaster preparation were found to be complementary rather than independent of men's. Nonetheless, women vary from males in that they save money, create portable stoves, and stockpile dry goods, molasses, biscuits, and critical medications.

If relief and reconstruction projects are going to minimize, rather than recreate, people's risk in future natural calamities, then they must understand the core reasons of gendered vulnerability. While highly vulnerable girls and women may not be considered in disaster plans, they still have critical needs, such as accessible and safe evacuation space; equitable access to food, clothing, and other relief goods; transportation assistance and emergency communication in community languages; childcare and other services supporting women's long-term care of surviving dependents; reproductive health care and gender-sensitive menta.

Disaster management in Bangladesh has been hampered by bureaucratic politics, which has reduced transparency within the country's Disaster Management Institutions and contributed to widespread failures in response. Among the many reasons that have complicated, the disaster risk reduction efforts of the Bangladeshi government are nepotism and clientelism. It would have been possible to replicate the successful—but isolated—initiatives undertaken by different stakeholders if the international humanitarian community had taken a more coordinated approach, particularly at the outset of the response, as well as between the international community and the government of Bangladesh. In spite of the lack of timely and effective coordination, both public and commercial organizations may successfully implement DRR programs without doubling down on resources.

References

1. Disaster Management Bureau (2004) Flood 2004: loss and damage. DMB, Dhaka
2. Disaster Management Bureau and Economic Relations Division (2005) Flood 2004 & its Impact on economy. ERD, Dhaka
3. Directorate of Relief and Rehabilitation (2007) Flood 2007: impact on lives and livelihood. DRR, Dhaka
4. DFID (2007) Flood causing huge devastation in Bangaldesh. DFID, Dhaka

5. Disaster Management Bureau (2007) Cyclone Sidr 2007 and its cuases and impact. DMB, Dhaka
6. Centre for Policy Dialogue (2008) Disaster and economic losses for Bangladesh. CPD, Dhaka
7. World Bank (2009) Economic impact of Cyclone Sidr and Flood in Bangladesh. World Bank, Dhaka
8. Directorate of Relief and Rehabilitation (2009) Cyclone Aila 2009: impact on lives and livelihood. DRR, Dhaka
9. Wazed M (2011) Disaster management in Bangladesh after 2007. MoFDM, Dhaka
10. The Daily Star (2004) Flood 2004: unbearable suffering of the poorest in North Bengal. The Daily Star, Dhaka
11. The Daily Star (2004) Budget and natural disasters in Bangladesh. The Daily Star, Dhaka
12. USAID and World Bank (2008) Cyclone Sidr making huge impact on economy. World Bank Office, Dhaka
13. Directorate of Relief and Rehabilitation (2008) Caretaker government & relief assistance to Sidr. DRR, Dhaka
14. The Daily Star (2008) Sidr-Barrier to overcome growth of national economy. The Daily Star, Dhaka
15. The Daily Prothom Alo (2009) Cyclone Aila: new AL government in action. The Prothom Alo, Dhaka
16. Directorate of Relief and Rehabilitation (2010) Huge allocation by the AL government for Aila recovery. DRR, Dhaka
17. CDMP (2010) Revised SOD: directives of AL government for better coordination. MoFDM, Dhaka
18. Cyclone Preparedness Programme (2010) Government action plan for CPP volunteer training. BDRCS, Dhaka
19. Disaster Management Bureau (2010) SSNP for disaster victims. DMB, Dhaka
20. Ministry of Food and Disaster Management (2011) DM plan for DRR in Bangladesh. MoFDM, Dhaka
21. Islam, Mohammad T. (2014) Climate change diplomacy—Apparatus for climate change mitigation and adaptation: a reflection in the context of Bangladesh. London, British Journal of Environment and Climate Change
22. Islam MT (2018) How does local government cope with disaster in Bangladesh? LSE South Asia Centre, London
23. CARE (2007) Cyclone SIDR response programme completion report. CARE, Dhaka
24. Islam MT (2018) Climate negotiations: how does Bangladesh fare? LSE South Asia Centre, London
25. UNRCO and MoFDM (2012) LCG for disaster risk reduction in Bangladesh. UNRCO, Dhaka
26. Economic Relations Divison and MoFDM (2012) Coordination of development partners for DRR in Bangladesh. ERD, Dhaka
27. Disaster Management Bureau (2011) Capacity development for DM officers at field level. DMB, Dhaka
28. SDMC (2003) Gujrat Earthquake 2001: wave of reconstruction. SDMC, New Delhi
29. Islam MT (2018) Despite legislative and institutional arrangements, Bangladesh is struggling to improve environmental governance. LSE South Asia Centre, London
30. Department of Disaster Management (2012) Cyclone Shelter center: best examples of disaster loss mitigation in Bangladesh. DDM, Dhaka
31. DDM (2012) Urban disaster risks in Bangladesh. DDM, Dhaka
32. Department of Disaster Management (2014) Cyclone Mahasen: family housing reconstruction project. DDM, Dhaka
33. UNO (2015) SDGs dealing with disasters. UNO HQ, New York
34. The Daily Star (2011) Disaster and poverty in Bangladesh. The Daily Star, Dhaka
35. CEGIS (2009) State of river water flows in Bangladesh and its impact on disasters. CEGIS, Dhaka
36. IUCN (2010) Mangrove Forest—Threat to disaster. IUCN, Dhaka

Chapter 6
Coronavirus Disease is More Severe Than Any Form of Natural Disasters

6.1 Local Government in Public Awareness and Preventive Measures

The world's nations are making heroic efforts to contain the newly emerged worldwide problem of COVID-19. COVID-19 is a recently identified coronavirus that causes an infectious condition known as coronavirus disease. In most cases, those infected with the COVID-19 virus will have only mild to moderate respiratory symptoms and make a full recovery without any further medical intervention.[1] Those at a later age and those who already have a preexisting condition like high blood pressure, diabetes, asthma, or cancer are at a higher risk for developing a life-threatening illness. In-depth knowledge of the COVID-19 virus, the illness it produces, and its transmission dynamics is the greatest defense against and response to the epidemic. Avoid spreading germs by touching your face and instead washing your hands often or using an alcohol rub.[2]

You should also observe respiratory etiquette to prevent the transmission of the COVID-19 virus, which is communicated mostly by droplets of saliva or nasal discharge from an infected person's coughing or sneezing (e.g., by coughing into a flexed elbow). It's important to note that Bangladesh is just as susceptible to COVID-19 as any other country. However, this worldwide epidemic will provide a tremendous challenge, since it is already failing to provide basic healthcare facilities in one of the world's most populous nations. As it has done in the past when faced with natural calamities, Bangladesh is showing that it can react effectively.

The Directorate General of Health Services of the Ministry of Health and Welfare drafted the National Preparedness and Response Plan for COVID-19, Bangladesh (version-5) in March of 2020 with the goal of preventing and controlling the spread of COVID-19 in Bangladesh and minimizing the virus's negative effects on the

[1] To find out more, please visit https://en.wikipedia.org/wiki/Coronavirus.

[2] To find out more, please visit https://www.who.int/emergencies/diseases/novel-coronavirus-2019.

M. T. Islam, *Disaster, Governance and Development*, Disaster Studies and Management,
https://doi.org/10.1007/978-981-99-1631-3_6

country's health, welfare, and economy. The goal of the strategy is to prevent the illness from entering the nation and to stop or contain its spread inside the country if it is brought in. According to the prevalence of the COVID-19 infection in a given nation, six distinct levels of planning and response have been established.[3] In order to establish, modify, or upgrade the degree of reaction, a risk assessment must be performed.

By categorizing countries according to their amount of COVID-19 infection, we can better prepare for and prioritize our responses. In order to establish, modify, or upgrade the degree of reaction, a risk assessment must be performed. In level 1, no cases are present; in level 2, imported cases are present; in level 3, local transmission is restricted; and in level 4, local transmission is widespread. In level 5, transmission slows down, and in the last stages, the system begins to recover. Committees at all levels, from the national to the Upazila, will carry out the work outlined in the national plan. These committees will include representatives from the appropriate ministries as well as national and international organizations and development partners. Specifically, the strategy establishes mechanisms for building up surge capacity to handle patients, maintain critical services, and mitigate societal effects. Any new information, R&D breakthroughs, worldwide good practices, or revised recommendations from WHO will need a reevaluation of the response plan and activities to ensure the most effective use of money and human resources. The ability to detect and respond to outbreaks of disease is crucial for efforts to limit their spread.

When it comes to preventing and recovering from natural disasters, Bangladesh is now a model for the world. The government of Bangladesh deserves credit for its efforts to lessen the impact of natural disasters such as typhoons, floods, droughts, and earthquakes. Whether it's a man-made or natural catastrophe, the Department of Disaster Management, which is part of the Ministry of Disaster Management and Relief, is always ready to respond. Authorities are obligated to take preventative measures in the face of natural catastrophes according to the Standing Order on Disaster and Disaster Management Act of 2012. The government of Bangladesh should use its Disaster Management Capability to combat the coronavirus outbreak.[4]

The Bangladeshi Army is still out there helping the civilian authorities maintain social distance and strengthen coronavirus preventative measures. Government agencies are making a concerted effort to prohibit needless public gatherings in an effort to limit the spread of coronavirus. Low-income individuals who have already relocated to villages outside of Dhaka and other divisional cities should be provided with enough food to last a month if the "remain at home or home quarantine" policy is to be successful. Selecting low-income persons living in rural regions of the nation and providing them with food for a month may be done jointly by local government and local administration. Let's work out a plan to keep rural Bangladeshis at home

[3] To find out more, please visit https://reliefweb.int/sites/reliefweb.int/files/resources/nprp_covid-19_v6_18032020.pdf.

[4] To find out more, please visit https://modmr.gov.bd/.

as much as possible. Union Parishads, as we are all aware, consist of a total of 13 delegates and a chairman.

One village police officer is often assigned to each community, and one office bearer directs the activities of all the selected union members to ensure safety and harmony in rural areas. The local media in Bangladesh state that foreign nationals make up the vast majority of individuals observing the house quarantine in the country's many rural communities. They come from all over the globe, including places like Italy, Dubai, Kuwait, Saudi Arabia, the United States, the UK, Spain, France, and Oman. Meanwhile, indications indicate that many individuals subjected to a house quarantine are acting in defiance of official directives. Many individuals are relocating to new areas in disobedience of the orders. We are well aware of the importance of local government in ensuring the well-being of rural Bangladeshis. Monitoring the home quarantine by Union Parishad officials with the help of village police may stop the spread of the fatal coronavirus among villagers.

There are an estimated 16 million disabled individuals in Bangladesh (10% of the population); however, I prefer to refer to them as "differently able persons." Many of them live in rural areas of the nation. Complications with COVID-19 seem to be more common among persons with impairments or long-term health issues than in the general population. While the government of Bangladesh is required to provide services to people with disabilities, such services have been narrowly focused on metropolitan areas and the disability issue has not been truly incorporated into general development efforts. What we should all do to be safe and slow the spread of coronavirus is obvious at this point. However, even routine tasks like washing one's hands may be challenging for certain handicapped persons. Some of them have limited range of motion in their hands. Some of them have trouble getting soap from a dispenser or reaching a sink. Many persons with disabilities also lack the means to live completely independently. They need the physical presence of others and the assistance with private, daily activities. Therefore, in order to maintain their support system, they must request that others around them reduce their own levels of protective isolation. All caregivers, whether they are friends and family or professionals, are placed in difficult situations.

They have to follow the rules set down by their respective agencies, pay attention to the pleadings of public health officials, watch out for the welfare of their own families, and carry out their duties as caregivers. The government should prioritize the needs of persons who are handicapped or have a long-term illness when considering the effects of coronavirus. The government, in solidarity with those on low incomes, should provide the greatest assistance feasible via a need-based plan to counter the threat of the fatal coronavirus. I have no doubt that they have the warm backing of Prime Minister Sheikh Hasina. Prime Minister Sheikh Hasina says, "We should not consider people with disabilities as burdens on society," emphasizing the need for a shift in perspective and calling on like-minded non-governmental organizations and donor agencies to collaborate with the government to meet the pressing needs of the disabled community.

A disaster management committee should be formed at the Upazila Parishad and the Union Parishad to provide the best possible assistance in halting the spread of the deadly coronavirus disease. This disease is much more dangerous than any natural disaster, and it might even become worse.

Similarly, the Union Parishad's standing committee for health services has to be functional if it is to fulfill its role as a service delivery organization (the extent to which this is the case is up to debate).

Chapter 7
Conclusion

7.1 Introduction

To successfully move DRR into the realm of development, strong leadership and a shared vision of DRR as developmental rather than catastrophe management are essential. To reduce disaster risk and help vulnerable people manage their lives and livelihoods in a more secure and less risky way, DRR must be mainstreamed, meaning that all stakeholders, both state and non-state, must embrace DRR into their respective development plans. Government, in addition to non-state actors, must act on the provisions of the Disaster Management Act and SOD to mainstream DRR into development planning by integrating DRR into all relevant ministry and departmental development plans. Disaster risk reduction is a crucial development problem that has to be assigned to a variety of plans and strategies, such as the Annual Development Plan and the Poverty Reduction Strategy Paper.

Humanitarian organizations struggled to coordinate their efforts after a tragedy because they all had their own plans of action, with goals that were too dissimilar to work together effectively. Fewer storm shelters were available than needed. On top of that, the Department of Catastrophe Management has trouble getting enough trained staff mobilized in the event of a disaster.

Furthermore, the infrastructure for data storage, compilation, and analysis was insufficient, and local disaster management committees lacked awareness and expertise concerning requirement assessment. Both the District and Upazila Disaster Management Committees lacked political leadership. The role of communicating with local disaster management committees was given to permanent representatives (Bangladeshi civil servants) of the central government. The Upazila Disaster Management Committee is headed by the Upazila Nirbahi Officer, while the District Disaster Management Committee is headed by the Deputy Commissioner.

The Disaster Standing Orders delegated power for coordinating and managing disaster risk reduction to individuals who are not professionals in disaster management. Political leadership at the local level is not engaged in catastrophe risk reduction, which means that the interests of the people and a feeling of responsibility

M. T. Islam, *Disaster, Governance and Development*, Disaster Studies and Management,
https://doi.org/10.1007/978-981-99-1631-3_7

are not represented. Catastrophe victims in Bangladesh have no idea what kinds of disaster risk reduction activities are in the works at the Upazila or district levels. However, the Union Chairman's role as chairman of the Union Disaster Management Committee is just on paper, and the committee's other members lack the necessary training to effectively carry out their duties.

7.2 Limitations of Policy Framework in Bangladesh

Increased requests from both the people of disaster-prone Bangladesh and the international community led to the development of a policy framework to institutionalize the government's efforts to reduce disaster risk for a safer community in Bangladesh. If this is the case, the policy framework may either restate tried-and-true tactics or provide a novel, far superior strategy that was overlooked by the relevant government. Throughout the policy framework, but particularly at this last step, the participation of key governmental and non-profit organizations is essential in order to give the required backing and incentive for such communities to execute, monitor, and assess such a plan.

Such a policy plan may be difficult to execute without encouragement in the form of follow-up activities and the availability of support, not because the communities are incompetent, but rather because of an engrained "handout" culture in which aid is expected. This procedure, however, may assist in overcoming such a culture since it requires the multi-stakeholder group to put into action a plan for which they are ultimately responsible. The findings of this research paper suggest that the government of Bangladesh should negotiate with the relevant stakeholders, including the community, on how to analyze the efficacy of each strategy, both in the past and in the present, and review the priority intrinsic components identified and all of the global and regional drivers of DRR.

7.3 Overall Challenges

It is challenging to include long-term DRR issues in Bangladesh because of the country's high catastrophe risk and the fact that post-disaster rehabilitation often prioritizes urgent requirements. It is true that the government of Bangladesh has established suitable legal provisions for disaster risk management, such as the mainstreaming of disaster risk reduction into development, but the degree of enforcement is insufficient.

The Department of Disaster Management operates under the Ministry of Disaster Management and Relief, although disaster risk reduction is a systemic problem that requires the attention of all government institutions. There currently isn't a central government body in charge of setting broad policies for disaster risk management, supervising their execution, and lobbying for disaster risk reduction to be included into overall development plans.

To hasten the transition from a culture of relief to one based on DRR, the government of Bangladesh must overcome the ongoing challenge of developing national policy and guidelines for integrating DRR into emergency preparedness, response, and recovery programs in light of DRR policies at the national level. However, the government of Bangladesh still needs to learn more about disaster risk reduction and associated measures. The country's disaster response infrastructure is set up to handle one big catastrophe every year. While progress has been made in disaster risk reduction (DRR), Bangladesh has experienced two national-scale catastrophes (in 2007 and 2009) that need additional confirmation of the underlying premise, especially in the context of climate change.

Furthermore, there is a lack of a systematic strategy to documenting DRR activities in communities for the purpose of duplicating and expanding such practices nationally. The local DMCs lack the power and resources necessary to design and execute DRR, and we've seen that financing for DRR at the community level is both scarce and unstable in Bangladesh.

7.4 Recommendation for Future Research

The purpose of this research was to better understand the DRR policy framework implemented by the Bangladeshi government. The researcher's focus is on the public policy implications of DRR, thus that is the angle she selected to study the topic from. While there is a wealth of DRR literature that explains and benefits policymakers' viewpoints, there has been no prior research using this perspective, which leaves opportunity for enhanced policy formation. However, additional study is needed, and the interaction between policymakers and implementers should be taken into account, to provide a more precise overall understanding of the full public policy framework in disaster risk reduction. An improved comprehension of the government of Bangladesh's DRR policy development and implementation procedures will result from this. Bangladesh's policies for preventing and responding to natural disasters might benefit from additional study.

Multiple more concepts emerged throughout the research and analysis phases that bear further exploration. The community's perspective on the government of Bangladesh's efforts to incorporate DRR concepts into its development projects is an intriguing area for further research. Because DRR lessens the financial burden of disaster response, this research would explain how DRR was included into the planning of an area's initial development scheme and how that plan evolved into current best practice.

7.5 Way Forward for Improvement of Policy Framework

Many efforts are under progress in Bangladesh to begin the process of mainstreaming catastrophe risk reduction into development. The following recommendations outline some of the activities that should be done to raise awareness of the need of mainstreaming and to finish setting up a suitable enabling environment.

The government of Bangladesh has implemented appropriate legal frameworks for disaster risk management, which include the mainstreaming of disaster risk reduction into development. These arrangements must be competently enforced if Bangladesh is to keep the momentum it has acquired. The Ministry of Disaster Management and Relief is in charge of the Department of Disaster Management; however, disaster risk reduction is a systemic problem that has to be "owned" by all government institutions. However, a centralized national body is necessary to take the lead on disaster risk management, design comprehensive policies, oversee their implementation, and push for disaster risk reduction to be included into planning and building. In disaster-prone nations like Bangladesh, a systematic, "joined-up" approach to disaster risk reduction and mainstreaming is an essential part of development initiatives.

Because industrialized nations have a moral duty to provide financial and technical assistance to developing nations in their attempts to reduce catastrophe risk, the developing world is demanding climate justice. The danger of climate-related disasters may be mitigated, but only if diplomatic efforts are ramped up to win over major backing from affluent nations.

Catastrophes may be avoided, and danger can be nullified. Disaster vulnerability is reduced, and disaster resilience is built at all levels via the coordinated, concerted, and deliberate actions of all stakeholders in all spheres of society. Our first line of defense against climate change is national government-led disaster risk reduction, which is also an effective strategy for promoting climate change adaptation at the international, national, and local levels, in particular through the practice of "mainstreaming" disaster risk reduction into all development efforts.

Results show that national governments have a strong commitment to disaster risk reduction, and this momentum should be maintained and perhaps increased. Get ready for calamities that might strike at any time by stocking up on supplies, logistics, and skilled personnel. To create a national strategy and standards for integrating DRR into disaster planning, response, and recovery programmes so that damaged areas may be rebuilt more quickly and with less disruption. DRR must be integrated into disaster response, preparation, and recovery; hence, it is imperative that the proper institutional structure be put in place to do so.

In order to ensure that government personnel have the skills necessary to carry out the DRR recommendations, a thorough training program should be maintained. The Ministry of Disaster Management and Relief has a moral obligation and an important leadership mandate to establish policy environments conducive to disaster risk reduction, to introduce effective approaches for disaster-resilient development, and to ensure that the required changes are implemented.

Our government has a responsibility to the people of Bangladesh to perform its oversight role in the formulation of national policies and their execution and to guarantee the efficient and accountable use of public monies in furtherance of the country's disaster risk reduction program. Information on how DRR may save money in the long run compared to responding to and recovering from disasters has to be shared extensively.

There is a pressing need for the government to raise all lawmakers' awareness of the role catastrophe risk reduction plays in safeguarding the populace and bolstering growth. The government should be committed to shifting national and global disaster risk reduction goals from "reduction" to "elimination" and to promoting disaster prevention with a "zero tolerance" approach to disaster losses as a frame of mind and strategy for global, national, and local development action.

To create a sustainable and fair world and a disaster-resistant human civilization for future generations, governments must reevaluate current development patterns while taking into account the socioeconomic standards and quality of life for our people. When it comes to development concerns like climate change, poverty, gender, education, public health, maternity and childcare, the environment, and infrastructure, among others, our ability to comprehend the connections between these problems and disaster risk reduction is essential. Both the planning and implementation of the DRR programme on the ground and the mobilization of resources for it must follow the principles of disaster governance. For the MDGS to be met by 2015, we must become change agents that look above partisan lines to promote disaster risk reduction and build synergies with adaptation to climate change.

Integrate measures taken to reduce the likelihood of disasters occurring into overall development strategies, plans, and activities by adopting a national strategy and putting in place the appropriate institutional structures and procedures. If you're concerned about long-term growth, it's important to redirect funds from expensive disaster relief and recovery operations to inexpensive catastrophe risk mitigation strategies. The danger of natural disasters may be reduced by increasing national and local budget allocations and development financing. Boost regional development investments and resource sharing via political commitment to catastrophe risk reduction. At several tiers, disaster management coordination committees need political leadership. Those in political leadership positions, with assistance from their respective federal and state governments, should serve as the hubs of coordination.

To ensure that the next generation of leaders in the country understands the importance of disaster risk reduction (DRR) to the country's long-term economic and social growth, it must be incorporated into the country's primary educational systems, local government programs, and national and local youth programs. Encourage international donors and financial institutions to back creative financing methods like debt swaps for initiatives to reduce the likelihood of catastrophe occurring in the first place. Make advantage of the donor-government dialogues that precede the development of country assistance plans to create programs that take risks at both the national and subnational levels into account. The political government must provide strong and capable leadership to guarantee donor responsibility.

Actions and initiatives to reduce catastrophe risk should be situated within this plan rather being taken in isolation. The policy should also include particular entry points and processes for incorporating disaster risk reduction into the overall development agenda and individual development projects. Assess the integration of humanitarian and development policies and regional organizations, and work to enhance mechanisms for the gathering and analysis of catastrophes and their direct linkages to ways to cost-effective hazard mitigation.

Disaster management is still widely alluded to in department circulars and executive orders, and it is the duty of individual line ministries, departments, and local governments to execute it within their spheres of responsibility. Unfortunately, few LGUs have produced a disaster management action plan, and the quality of those that have varies widely. In addition, SOD notes that the risk reduction and mainstreaming skills and competencies of many disaster management committees are severely lacking, since their meetings tend to be called solely in reaction to crises rather than on a regular basis to address continuing risk reduction activities. Disaster management committees at both the district and Upazila levels follow the same protocol. To effectively promote a culture of disaster risk reduction, local government units and local-level disaster management committees must become more organized and effective.

To keep everyone focused, DRR needs a distinct identity that sets it apart from traditional disaster management focused on providing aid. DRR's identity should be shaped in conversation with disaster management and development stakeholders from across all policy sectors with an interest in development or risk reduction, so that it is appropriate to national contexts. Those working on post-disaster rebuilding efforts should be required by law to consider catastrophe risk reduction while making decisions.

Larger coalitions between civil society and policymakers may include grassroots actors like community-based groups to better mobilize local resources and public opinion in support of reform. Advocacy still relies on the participation of civil society actors because they are a means by which grassroots perspectives can be shared, popular views can be represented, and public legitimacy and supervision can be fostered.

DRR mainstreaming is a method of governance that permits the methodical absorption of DRR issues into all relevant development arenas. In other words, catastrophe risk reduction may be institutionalized as a guiding principle of sustainable development if governance systems are responsive, responsible, transparent, and efficient. Therefore, in disaster-prone countries, it is essential that: (a) underlying risk factors are continuously considered in all relevant sectors; and (b) risk reduction standards and measures are integrated into the planning and delivery of fundamental development services and processes like education, environment, and health.

To put the law into effect and to provide coordination and monitoring methods and arrangements, a disaster risk management plan that involves all levels of government, the corporate sector, local communities, and civil society is necessary. Actions and initiatives to reduce catastrophe risk should be situated within this plan rather being taken in isolation. The policy should also include particular entry points and processes for incorporating disaster risk reduction into the overall development agenda and individual development projects.

When it comes to land use planning for disaster risk reduction, determining the right criteria is the biggest challenge. All relevant factors, including social, cultural, geographical, political, and economic ones, must be identified in order to arrive at a range for the relevant criteria. It's possible that the planning will fall through if any of these conditions isn't paid enough attention to. Executive laws controlling the physical form and quality of the urban environment and intervening tactics in urban fabrics are effective elements in making the land use plan strategic. Planning for the use of land may be an effective risk reduction strategy since it can lessen exposure to many different types of natural and man-made disasters.

The development of a "National Land Use Policy" is essential for effective land use planning taking DRR into account. What people desire or believe they ought to have, and how they see government's role in facilitating more efficient use of land and water, set the tone for land use policy's overarching goals. Only when individuals are unhappy with the status quo or have an idea for a better use of land and water do they want to make changes.

Measures to reduce the danger of natural disasters that don't involve building repairs need to be bolstered. More shelters need to be erected in cyclone-prone locations. Offering educational and counseling services in public forums like seminars and conversations may be quite effective. Regional collaboration should be explored for international river basin development. We need to install an accessible communication network in high-risk locations and keep a careful eye on them. ICT can assist ensure the timely and effective rollout of the warning system.

So that the relevant entities designated for disaster management may be mobilized swiftly, it is important to build a suitable program for disaster prediction and preparation. For better cooperation, a joint exercise involving civilians and the military should be planned. We have solid precedent in this respect, since the armed forces have historically been called upon to assist civil government in conducting combined and enormous operations for search-and-rescue and relief operations in the aftermath of catastrophic events.

For example, as governmental leadership and backing in Bangladesh weakened, reform momentum slowed, stakeholder buy-in decreased, and those participating felt disheartened. An influential policy advocate who is also technically savvy, skilled, dedicated, innovative, and well-respected may do wonders for a reform's chances of success. Policy champions, ideally holding senior political or administrative positions, should be conversant with and adept at navigating both formal and informal institutional settings. If legislation is to be the primary driver of cross-sector integration, it should reside at the highest level of executive power.

Every effort should be made to ensure that new laws can be effectively implemented and monitored within the framework of current government planning cycles and processes. Any plan to improve disaster preparedness must take into account the many different sectoral policies and programs already in place. This lessens the risk of resource rivalry and minimizes the perception that DRR legislation and implementation are too expensive. When it comes to enforcing DRR laws and working with line ministries and other stakeholders to implement risk reduction measures, well-established national platforms may play a significant role. There is a need for regional organizations to aid in the creation and enforcement of national laws. Therefore, it is crucial that DRR laws be in line with global best practice. People at the center of the reform effort, such national professional organizations and educational institutions with a focus on disaster risk management, should have ties to groups working to reduce the likelihood of catastrophic events.

The DRR Act must result in administrative change at the lowest level and the deployment of financial resources in local-level activities. It is crucial to provide local governments with resources and expertise in addition to the duty for risk reduction, since this is where national law is really implemented.

Pragmatism dictates that Bangladesh uses a multi-level approach to DRR. Collaborating between the public and commercial sectors might lead to an increase in catastrophe risk reduction initiatives, which is everyone's business. We are all aware that multi-level governance is considered a novel method to governance. What we seem to be seeing is the formalized collaboration between different types of government in a given area or between different groups inside a single country, all with the same objective in mind and intending to work together to make it a reality.

As a first step toward this end, multi-level governance sheds light on the interplay between national and regional governments and other public and private players in the formulation and implementation of policies at various tiers of government, from the global to the neighborhood. No matter the political structure, multi-level governance requires the implementation of mechanisms for both vertical and horizontal collaboration in order to reduce the policy "gaps" across different tiers of government. The following diagram may be useful in the context of Bangladesh for encouraging collaboration among important players in the context of disaster risk reduction (Fig. 7.1):

Create a norm and set up a system to record DRR practices in communities so that they may be replicated and expanded throughout the country to guarantee disaster management services at worldwide levels of quality. Rebuilding devastated areas via the methodical inclusion of risk reduction strategies into disaster planning, response, and recovery plans.

Bangladesh's policy framework has shown a novel approach to disaster risk reduction, one that allows communities to identify and explore viable solutions to their susceptibility to environmental hazards. The framework's conceptual structure does not effectively reflect the utility or the possible solutions to promote catastrophe risk reduction as a whole, despite the fact that it has been created in conjunction with key parties. Recognition by connected stakeholders that enforcement of applicable rules throughout the year is vital is the first step toward execution of the disaster risk reduction strategy. Policy framework without enforcement is a waste of time since it will lead to top-down solutions being imposed on bottom-up problems.

Only through mutually respectful and open discussion between stakeholders and the community at large can an integrated strategy be developed, and a policy framework implemented. Communities are better able to understand the value of disaster

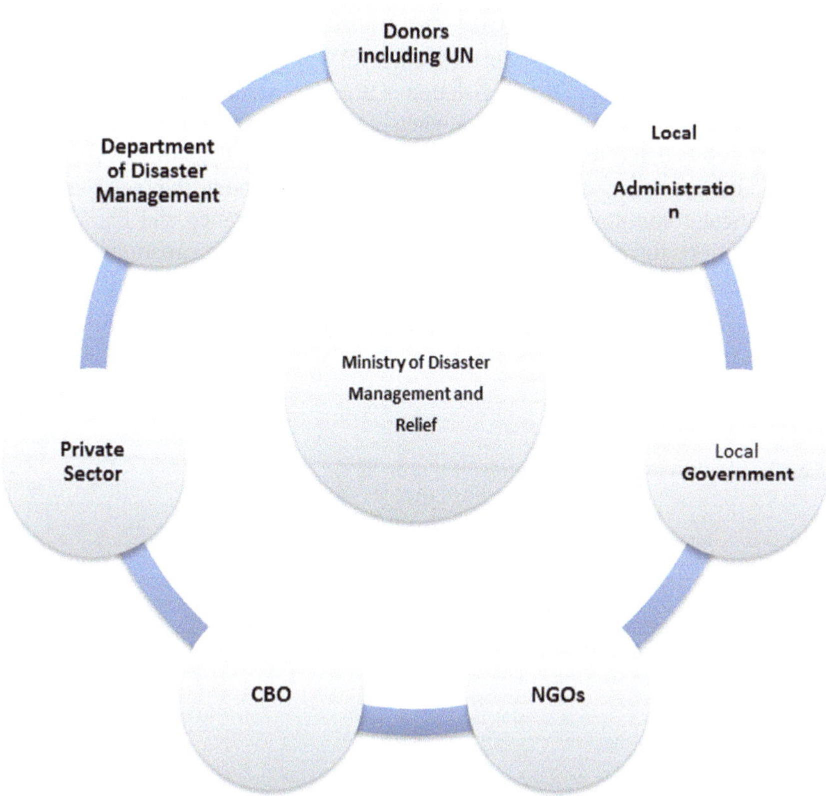

Fig. 7.1 Relevant actors for disaster risk reduction

risk reduction programs and how to incorporate them to lessen vulnerability when they can trust and communicate with those responsible for making policy decisions.

The policy framework creates an environment where the two bodies of knowledge may be successfully integrated at the local level. This is neither a top-down or bottom-up endeavor, but rather a community-wide and stakeholder-wide effort.

Comprehensive analysis of disaster risk management financing, including public sector spending on DRR, relief, and response; assessment to identify key gaps in DRR capacity; and incorporation of relevant DRR activities into budgeted projects and development plans at the appropriate level are all necessary for effective capacity development, advocacy, and programming. Advocacy, education, and public awareness on disaster risk reduction are essential for maximizing the advantages of the policy framework established by the government of Bangladesh.

According to the aforementioned suggestion, for DRR legislation to be effective in influencing the most vulnerable populations to alter their behavior, it must include provisions that clearly define the outcomes that must be achieved at the community level, lead to changes at the lowest administrative level, and result in the allocation of resources to activities at the local level. The degree to which DRR law in Bangladesh leads to good practice and change at the local level will be decisive in determining the efficacy of the legislation. DRR legislation is urgently needed, but there are several obstacles and possibilities to consider when crafting and enforcing such a law on a national level so that it reaches those who are most at risk.

Last but not least, widespread agreement exists about the need of high-level political commitment armed with appropriate experience and knowledge in bolstering efforts toward catastrophe risk reduction and its mainstreaming into wider development.

Bibliography

1. Ahmed KI (2001) Participatory action research on building-for-safety options for low-income rural housing in flood-prone areas. University of Exeter, London
2. ADPC (2008) Community-based disaster risk management field practitioners' handbook. ADPC, Bangkok
3. ADB (1992) Disaster mitigation in Asia and the Pacific. ADB, Manila
4. Ara S (1998) Participatory monitoring and evaluation of flood-proofing pilot project of CARE-Bangladesh. IIED, London
5. Anderson JR (1990) The adaptive character of thought. Erlbaum, Hillsdale
6. Arrow KJ (1982) Risk perception in psychology and economics. Economic Inquiry, London
7. Black R (2008) Demographics and climate change: future trends and their policy implications for Migration. Development Research Centre on Migration, Globalisation and Poverty, University of Sussex, Brighton
8. Blakie P (1994) At risk: natural hazards, people's vulnerability and disasters. London, Routledge
9. Benouiar D (2003) Contribution to the UN-ISDR's online discussion on the draft framework to guide and monitor disaster risk reduction. Geneva UNISDR Office
10. Brokensha D, Warren D, Werner O (1980) Indigenous knowledge systems and development. University Press of America, Washington, DC
11. Cardona OD (2004) The need for rethinking the concepts of vulnerability and risk from a holistic perspective: a necessary review and criticism for effective risk management. Earthscan Publishers, London
12. Chaskin R, Brown P, Venkatesh S, Vidal A (2001) Building community capacity. Aldine de Gruyter, New York
13. Collier P, Hoeffler A (2002) Aid, policy, and growth in post-conflict societies. World Bank HQ, Washington, DC
14. Collier P, Hoeffler A (2002) Aid, policy and peace: reducing the risks of civil conflict. World Bank HQ, Washington, DC
15. Centre for Research on the Epidemiology of Disaster (2010) The international disaster database. Catholic University Press, Belgium
16. Carter WN (1991) Disaster management: a disaster manager's handbook. ADB, Manila
17. Crawford A (1997) The local governance of crime: appeals to community and partnership. Oxford University Press, UK
18. CDB and CARICOM Secretariat (2004) Sourcebook on the integration of natural hazards into environmental impact assessment. Caribbean Development Bank and Caribbean Community Secretariat, Barbados
19. Campitelli G, Gobet F (2004) Adaptive expert decision making: skilled chess players search more and deeper. International Computer Games Association, ICGA, Netherlands

M. T. Islam, *Disaster, Governance and Development*, Disaster Studies and Management,
https://doi.org/10.1007/978-981-99-1631-3

20. Chaiken S, Trope J (1999) Dual-process theories in social psychology. Guilford Press, New York
21. DFID (2006) Reducing the risk of disasters—helping to achieve sustainable poverty reduction in a vulnerable world, UK DFID London
22. Dye TR (1976) Policy analysis. University of Alabama Press, USA
23. Dekens J (2007) Local knowledge on disaster preparedness: a framework for data collection and analysis. Sustainable Mountain Development Programme, Nepal
24. Ericksen NJ, Ahmad QK, Chowdhury AR (1997) Socio-economic implications of climate change for Bangladesh. Bangladesh Unnayan Parishad, Dhaka
25. Ericsson KA, Charness N, Feltovich PJ, Hoffman RR (2006) The Cambridge handbook of expertise and expert performance. Cambridge University Press, New York
26. Holloway A, Pelling M (2006) Legislation for mainstreaming disaster risk reduction. Tearfund, Teddington
27. Hewitt K (1999) Regions of risk: a geographical introduction to disasters. Longman, Harlow, Essex
28. Haque CE, Blair D (1992) Vulnerability to tropical cyclones: evidence from the April 1991 cyclone in coastal Bangladesh. Disasters Management Bureau, Dhaka
29. Hurlbert M (2018) Adaptive governance of disaster: drought and flood in rural areas. Springer International Publishing, Switzerland
30. IDB (2006) Draft disaster risk management policy. KSA, IDB
31. Islam MT (2014) Climate change diplomacy—apparatus for climate change mitigation and adaptation: a reflection in the context of Bangladesh. Lond Br J Environ Clim Change
32. Islam MT (2017) Understanding the effectiveness of Union Parishad Standing Committee: a perspective on Bangladesh. LSE South Asia Centre, London
33. Islam MT (2018) How does local government cope with disaster in Bangladesh? LSE South Asia Centre, London
34. Islam MT (2019) Rural dispute resolution in Bangladesh: how do village courts safeguard justice? Contemporary South Asia, London
35. Islam MT (2016) Despite constitutional guarantees, Bangladesh is failing to deliver adequate healthcare to rural citizens. LSE South Asia Centre, London
36. Islam MT (2016) Despite constitutional guarantees, Bangladesh is failing to deliver adequate healthcare to rural citizens. LSE South Asia Centre, London
37. Islam MT (2020) Bangladesh: towards the localisation of the sustainable development goals. LSE South Asia Centre, London
38. Islam MT (2020) Towards the localisation of the SDGs. The Daily Star, Dhaka
39. Islam MT (2020) Combating COVID-19 in rural Bangladesh: the role of the local government. NUS Institute of South Asian Studies, Singapore
40. Islam MT (2020) Bangladesh. In: Quest of a decentralised local representative bodies. GIZ Urbanet, Germany
41. Islam MT (2020) State of human security in Rural Bangladesh amidst COVID Pandemic. Khabarhub, Kathmandu
42. Islam MT (2020) How UN defends global peace, prosperity amid COVID-19 pandemic? Khabarhub, Kathmandu
43. Islam MT (2020) Localizing SDGs is must for inclusive rural development in Bangladesh. Khabarhub, Kathmandu
44. IDNDR (2000) Natural disasters: strategies for mitigation and disaster response. IDNDR, Germany
45. Kratt P (2005) Reducing the risk of disasters: SIDA's effort to reduce poor people's vulnerability to hazards. Swedish International Development Cooperation Agency, Stockholm
46. Karim MF, Mimura N (2008) Impacts of climate change and sea level rise on cyclonic storm surge floods in Bangladesh. Global Environmental Change, London
47. Kendra JM, Wachtendorf T (2003) Elements of resilience after the world trade center disaster. Reconstituting New York City's Emergency Operations Center, New York

48. Shah Alam Khan M (2008) Disaster preparedness for sustainable development in Bangladesh. Disaster Prevention and Management, Dhaka
49. Krüger F, Bankoff G, Cannon T, Orlowski B, Schipper E, Lisa F (2015) Cultures and disasters: understanding cultural framings in disaster risk reduction. Routledge, UK
50. Lewis L (1999) Development in disaster prone places: studies of vulnerability. Intermediate Technology Publications, London
51. Lowis TJ (1985) The state in politics. University of California Press, USA
52. Mamun MZ (1996) Awareness, preparedness and adjustment measures of river-bank erosion-prone people: a case study. Dhaka
53. Medina JC (1992) Community awareness and participation programme in disaster preparedness and rural development. UNCRD, Nagoya
54. Miller MA, Douglass M (2016) Disaster governance in urbanising Asia. Springer International Publishing, Switzerland
55. Office of the United Nations Disaster Relief Coordinator (1984) Preparedness aspects: disaster prevention and mitigation. United Nations HQ, New York
56. Pelling M (2003) The vulnerability of cities: natural disasters and social resilience. Earthscan, London
57. Peters K (2018) Accelerating Sendai framework implementation in Asia. ODI, London
58. Shah HC (2002) Beyond earthquake loss estimation technologies: strategies for risk reduction. Bucharest Press, Rumania
59. Schipper L, Pelling M (2006) Disaster risk, climate change and international development: scope for, and challenges to integration. Action Aid HQ, UK
60. Shea E (2003) Living with a climate in transition: pacific communities plan for today and tomorrow. East-West Centre, Honolulu
61. The Government of Bangladesh (2008) National Disaster Management Plan (Draft). MoDMR, Dhaka
62. The Government of Bangladesh (2008) Disaster management act 2012. MoDMR, Dhaka
63. The Government of Bangladesh (2008) The standing orders on disaster (revised). DMB, Dhaka
64. UNISDR (2004) Living with risk: a global review of disaster reduction initiatives. UNISDR, Geneva
65. UNISDR (2004) Terminology: basic terms of disaster risk reduction. UNISDR, Geneva
66. USAID (2007) How resilient is your costal community? A guide for evaluating coastal community resilience to tsunamis and other hazards. USAID, Bangkok
67. UN (2002) Living with risk—a global review of disaster reduction initiatives. United Nations, Geneva
68. UNDP (2004) Reducing disaster risk—a challenge for development, United Nations Development Programme. UNDP HQ, New York
69. UNDP (1997) The shrinking state-governance & sustainable human development. United Nations Development Programme, UNDP HQ, New York
70. Varma SP (1975) Modern political theory. Vikas Publishers, Delhi
71. World Bank (2002) World development through Building Institutions for Markets. Oxford University Press, Washington, DC
72. Wisner B (2004) At risk: natural hazards, people's vulnerability and disasters. Routledge, London
73. World Bank (2006) Hazards of nature, risks to development—an IEG evaluation of World Bank: assistance for natural disasters. World Bank HQ, Washington, DC
74. World Bank and UNISDR (1992) Global facility for disaster reduction and recovery: a partnership for mainstreaming disaster mitigation in poverty reduction strategies. UNISDR, Geneva

MIX
Papier aus verantwortungsvollen Quellen
Paper from responsible sources
FSC® C105338

If you have any concerns about our products,
you can contact us on
ProductSafety@springernature.com

In case Publisher is established outside the EU,
the EU authorized representative is:
Springer Nature Customer Service Center GmbH
Europaplatz 3, 69115 Heidelberg, Germany

Printed by Libri Plureos GmbH
in Hamburg, Germany